Current Topics in Environmental Health and Preventive Medicine

Series Editor

Takemi Otsuki
Kawasaki Medical School
Kurashiki
Okayama, Japan

Current Topics in Environmental Health and Preventive Medicine, published in partnership with the Japanese Society of Hygiene, is designed to deliver well written volumes authored by experts from around the globe, covering the prevention and environmental health related to medical, biological, molecular biological, genetic, physical, psychosocial, chemical, and other environmental factors. The series will be a valuable resource to both new and established researchers, as well as students who are seeking comprehensive information on environmental health and health promotion.

More information about this series at http://www.springer.com/series/13556

Georg Steinhauser • Akio Koizumi
Katsumi Shozugawa
Editors

Nuclear Emergencies

A Holistic Approach to Preparedness and Response

Springer

Editors
Georg Steinhauser
Institute of Radioecology
and Radiation Protection
Leibniz Universität Hannover
Hannover
Germany

Katsumi Shozugawa
Department of General System Studies
Graduate School of Arts and Sciences
University of Tokyo
Meguro, Tokyo
Japan

Akio Koizumi
Department of Health
and Environmental Sciences
Graduate School of Medicine and Faculty
of Medicine
Kyoto University
Sakyo-ku, Kyoto
Japan

ISSN 2364-8333 ISSN 2364-8341 (electronic)
Current Topics in Environmental Health and Preventive Medicine
ISBN 978-981-13-8326-7 ISBN 978-981-13-8327-4 (eBook)
https://doi.org/10.1007/978-981-13-8327-4

This Springer imprint is published by the registered company Springer Nature Singapore Pte Ltd.
The registered company address is: 152 Beach Road, #21-01/04 Gateway East, Singapore 189721, Singapore

Nothing in life is to be feared, it is only to be understood. Now is the time to understand more, so that we may fear less.

Marie Skłodowska Curie (1867–1934)

Contents

About the Editors

Georg Steinhauser is professor of Radioecology at Leibniz University Hannover. After his graduation in chemistry (MSc) from the University of Vienna (2003), he received his PhD in radiochemistry from Vienna University of Technology in 2005. Supported by a Schrödinger-Fellowship, he joined Prof. Klapötke's group at Ludwig Maximilian University of Munich, Germany, for one year in 2007. Steinhauser's scientific roots are at Vienna University of Technology, where he worked for 10 years using the Atominstitut's TRIGA research reactor, before he was hired in 2013 by Colorado State University's Department of Environmental and Radiological Health Sciences (assistant professor of radiochemistry). Since 2013, he is member of the Radiation Protection Advisory Board of the Austrian Federal Ministry of Health (Strahlenschutzbeirat). In October 2015, he was hired by Leibniz University Hannover (Institute of Radioecology and Radiation Protection) to assume his current position. His main research focus is environmental radioactivity in Chernobyl and Fukushima as well as environmental nuclear forensics and inorganic chemistry of the f-block elements. Steinhauser has (co-)authored more than 100 scientific publications. Since 2016, he is editor of the Springer journal *Environmental Science and Pollution Research*. In 2017, Steinhauser organized the Nuclear Emergency Expert Meeting (NEXT) in Hannover.

Akio Koizumi is professor emeritus of Health and Environmental Sciences, Graduate School of Medicine, Kyoto University. After his graduation from the School of Medicine (MD) at Tohoku University (1978) and an internship at Tohoku Rosai Hospital, he received his PhD from Tohoku University in 1983. He had a postdoc career in Health and Environmental Sciences Laboratory at the Dow Chemical Company in Midland, MI, from 1983 to 1985 and then joined the University of California, Riverside, as a Research Toxicologist from 1985 to 1987. He took an associate professor position at the Department of Hygiene, Graduate School of Medicine, Akita University, in 1987 and was promoted to professor in 1993. In 2000, he moved to the Department of Health and Environmental Sciences, Graduate School of Medicine, Kyoto University. In March 2018, he retired from Kyoto University (Professor Emeritus) and became a Director in the Public Health Institute affiliated to Public Interest Association Kyoto Hokenkai. His main research focus has been health promotion and prevention of diseases. He has organized several research programs, investigating long-term health effects after the Fukushima Daiichi Nuclear Power Plant accident. Koizumi has (co-)authored more than 380 scientific publications with about 13,000 citations and was awarded with several famous prizes. He also plays Igo.

Katsumi Shozugawa is assistant professor of Radiochemistry and Environmental Chemistry at the University of Tokyo. In 2004, he joined Prof. Matsuo's group at the University of Tokyo and received his master's and doctoral degrees in the field of environmental chemistry. He had been engaged in activation analysis and chemical states analysis using research reactors (JRR-3M, JAEA) and accelerators (Photon Factory, KEK), respectively. After the Fukushima nuclear power plant accident (2011), he is engaged in the quantification of radioactive materials released from the Fukushima NPP, mainly in the evacuation zone. He is also energetically conducting radiation education. As of the end of 2018, he lectured for more than 9000 children, students, and parents.

Chapter 1
Introduction to Nuclear Emergencies

Georg Steinhauser, Akio Koizumi, and Katsumi Shozugawa

Abstract Nuclear emergencies exhibit an imminent threat to the fabric of society as they may cause severe actual damage or may be perceived as hazardous. A holistic approach is needed to assess past accidents as well as future accident scenarios. For this reason, education, science, and research are needed now as well as in the future for proper nuclear accident preparedness and response. This chapter outlines some basic definitions and explains the history and concept of this publication.

Keywords Nuclear emergency · Radiological emergency · Nuclear accident

While memories of the nuclear accidents at Chernobyl and Fukushima are still present in many people's mind, one has to admit that nuclear emergencies are not only about the past but also about the future. Nuclear emergency scenarios—hopefully on a much smaller scale and hopefully preventable—are likely to happen again at some point in the future. Without any doubt, however, fear of radiation will stick with mankind as long as nuclear technology is being used. For this reason, education, science, and research are needed now and will be needed in the future for the preparedness, mitigation, and response to possible future nuclear accidents.

G. Steinhauser (✉)
Leibniz Universität Hannover, Institute of Radioecology and Radiation Protection, Hannover, Germany
e-mail: steinhauser@irs.uni-hannover.de

A. Koizumi
Department of Health and Environmental Sciences, Graduate School of Medicine and Faculty of Medicine, Kyoto University, Sakyo-ku, Kyoto, Japan
e-mail: koizumi@kyoto-hokenkai.or.jp

K. Shozugawa
Department of General Systems Studies, Graduate School of Arts and Sciences, The University of Tokyo, Meguro, Tokyo, Japan
e-mail: cshozu@mail.ecc.u-tokyo.ac.jp

G. Steinhauser et al. (eds.), *Nuclear Emergencies*,
Current Topics in Environmental Health and Preventive Medicine,
https://doi.org/10.1007/978-981-13-8327-4_1

The International Atomic Energy Agency (IAEA) Glossary defines the term "emergency" as follows [1]:

> A non-routine situation that necessitates prompt action, primarily to mitigate a hazard or adverse consequences for human health and safety, quality of life, property or the environment. This includes nuclear and radiological emergencies and conventional emergencies such as fires, release of hazardous chemicals, storms or earthquakes. It includes situations for which prompt action is warranted to mitigate the effects of a perceived hazard.

In particular, a "nuclear or radiological emergency" is defined by IAEA as follows [1]:

> An emergency in which there is, or is perceived to be, a hazard due to:

a) The energy resulting from a nuclear chain reaction or from the decay of the products of a chain reaction; or
b) Radiation exposure.

> [Points (a) and (b) approximately represent nuclear and radiological emergencies, respectively. However, this is not an exact distinction.]

Since we do not attempt to further distinguish between the exact nature of an emergency or incident, we will use both terms (nuclear and radiological emergency, respectively) synonymously.

Nuclear emergencies include nuclear and radiological accidents, the explosion of a (military or improvised) nuclear device, and nuclear terrorism. All possible scenarios (whether accidental or intentional) are capable of creating physical harm and horror among the affected population. Thus, they are possibly threatening the fabric of our society and require the development of options for the avoidance, preparedness, mitigation, and response to such an event. By looking at the above definition of a "nuclear emergency," it becomes apparent that nuclear emergencies not only include scenarios that involve ionizing radiation, radioactive or nuclear materials in a way that they pose an *actual* threat to human or environmental health, but they also include scenarios where such adverse consequences are only *perceived* by the public or the affected or unaffected communities. This aspect, in our opinion, has not yet received sufficient attention in (nuclear) emergency preparedness and response operations and protocols. This facet, therefore, has been the key motivation for this book as an attempt to elucidate and evaluate strategies for preparedness and response to nuclear emergencies not only on a technical, but on a holistic level.

In addition to the radiological and technical aspects of a nuclear emergency, scientists, stakeholders, authorities, and the public need to take into account the role of the media and environmental organizations, cultural aspects of "fear," the need for public education at a young age. Proper preparedness and response plans include a health evaluation of an incidence that includes a health physical, radioecological, analytical and radioanalytical, sociological, psychological, and medical perspective. In order to discuss these aspects, a meeting at the expert level was organized, which was held from August 30 to September 1, 2017, at Herrenhausen Palace (Fig. 1.1), in Hannover, Germany: The Nuclear Emergency Expert Meeting 2017 (NEXT 2017) brought together 28 participants from 6 countries (Austria, France, Germany, Japan, Ukraine, and the USA, Fig. 1.2).

Fig. 1.1 Herrenhausen Palace, Hannover, Germany

Fig. 1.2 Participants of NEXT

This book, however, is not intended as the sole proceedings publication of this conference. The contributions in this book go far beyond any summary of the respective talks. They are intensely and thoroughly researched contributions that represent the state of the art in the field of nuclear emergency preparedness and response at a holistic level.

Acknowledgments The NEXT 2017 symposium received generous financial support from the Volkswagen Foundation.

Reference

1. IAEA. IAEA safety glossary. 2007. https://www-pub.iaea.org/MTCD/publications/PDF/Pub1290_web.pdf. Accessed Jan 2019.

Chapter 2
Two Major Nuclear Emergencies: A Comparison of Chernobyl and Fukushima

Georg Steinhauser

Abstract In this chapter, various aspects of the two major reactor accidents at Chernobyl (1986; Ukrainian SSR) and Fukushima (2011; Japan) are discussed and compared. Both accidents have been ranked at the maximal level of 7 ("Major Accident") at the International Nuclear and Radiological Event Scale (INES). The Chernobyl nuclear accident was caused by an unauthorized experiment in combination with design flaws of the RBMK reactor. The Fukushima nuclear accident was caused by a natural disaster (a tsunami that was triggered by an earthquake). Both accidents released radionuclides mostly of the volatile elements (Kr, Xe, I, Cs, Te), but Chernobyl also released significant amounts of less volatile radionuclides (Sr, Ru, lanthanides, actinides, etc.), mainly in the form of hot particles. Much larger areas have been contaminated by the Chernobyl accident than by Fukushima. The health effects due to the nuclear accidents have been much more severe for the residents of the Chernobyl area than for the residents of the Fukushima prefecture.

Keywords Nuclear accident · International Nuclear and Radiological Event Scale (INES) · Chernobyl · Fukushima · Radioactive pollution · Radionuclides Environmental release · Emergency response · Evacuation · Contaminated areas Health effects

G. Steinhauser (✉)
Leibniz Universität Hannover, Institute of Radioecology and Radiation Protection, Hannover, Germany
e-mail: steinhauser@irs.uni-hannover.de

G. Steinhauser et al. (eds.), *Nuclear Emergencies*,
Current Topics in Environmental Health and Preventive Medicine,
https://doi.org/10.1007/978-981-13-8327-4_2

2.1 Introduction

A nuclear emergency involves ionizing radiation or nuclear/radioactive materials in a harmful manner. In most cases, nuclear emergencies are associated with the release of radioactive substances, but there have also been accidents, in particular criticality accidents, where harmful amounts of ionizing radiation were released [1]. However, in this chapter, two of the major nuclear reactor accidents that involved the release of radionuclides, Chernobyl and Fukushima Daiichi, shall be introduced and briefly compared.

2.2 The International Nuclear and Radiological Event Scale (INES)

In order to classify nuclear emergencies, the International Atomic Energy Agency (IAEA) has established the *International Nuclear and Radiological Event Scale* (INES) that allows for the comparison of the severity of a nuclear emergency (Fig. 2.1). This comparability, however, is somewhat limited since INES ratings are not conducted by a central body (e.g., within the IAEA), but by the operators of a facility or a regulatory body of the respective country. Table 2.1 lists a selection of the major nuclear accidents and some of their characteristics, including their INES ratings. Since the INES had been introduced by the IAEA only in 1990, many accidents and incidents had to be rated retrospectively.

In the public perception, nuclear emergencies are all about nuclear reactor accidents. Indeed, the only two nuclear emergencies that scored at the maximum rating level (7 —"Major Accident") of the INES have been nuclear reactor accidents, namely Chernobyl in the Soviet Union (1986) and Fukushima in Japan (2011).

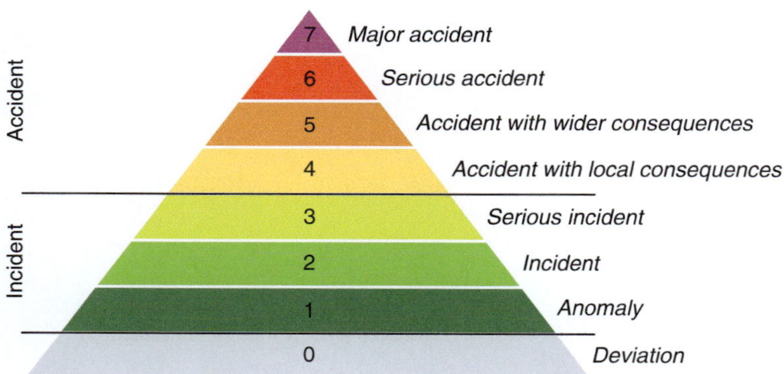

Fig. 2.1 The levels and definitions of the International Nuclear and Radiological Event Scale (INES)

Table 2.1 Overview and comparison of some major nuclear accidents in history

Location	Date	Type of accident	Total amounts of released radionuclides	Main released radionuclides	INES level
Chernobyl, USSR	1986/04/26	Reactor accident	5300 PBq[a]	^{133}Xe, ^{131,133}I, ^{132}Te, ^{134,136,137}Cs, ^{90}Sr, actinides	7
Fukushima Daiichi, Japan	2011/03/11	Reactor accident	520 (340–780) PBq[a]	^{133}Xe, ^{131,133}I, ^{132}Te, ^{134,136,137}Cs	7
Kyshtym, USSR	1957/09/29	Release during nuclear reprocessing	740 PBq	^{90}Sr, ^{95}Zr, ^{106}Ru, ^{144}Ce	6
Windscale, UK	1957/10/10	Reactor accident	18.6 PBq	^{133}Xe, ^{131,133}I, ^{132}Te, ^{137}Cs, ^{210}Po, ^{3}H	5
Three Mile Island, USA	1979/03/28	Reactor accident	1 TBq	^{133}Xe, ^{85}Kr, ^{131}I	5
Goiânia, Brazil	1987/09/13	Theft and unauthorized opening of a radioactive source	50.9 TBq	^{137}Cs	5

[a]Excluding noble gases

The total estimated releases, in activity,[1] were 5300 PBq (1 PBq = 10^{15} Bq) in case of the Chernobyl nuclear accident [2], and 520 (340–780) PBq for the Fukushima nuclear accident [3], which are both large amounts of radionuclides. However, when only based on the magnitude of its releases, the Fukushima nuclear accident has not been the second worst in human history: as illustrated in Table 2.1, it probably falls behind the Kyshtym nuclear accident in the Soviet Union (1957), which released an estimated amount of 740 PBq of (mostly medium- and long-lived) radionuclides in its vicinity, causing an even higher deposition densities [4]. The Kyshtym nuclear accident, however, was retrospectively rated "only" at the level 6 ("Serious Accident") of the INES scale. Here, we will focus only on Chernobyl and Fukushima. Plenty of literature exist on the other accidents, e.g., for Windscale [5], Kyshtym [4], Goiânia [6], or Three Mile Island [7].

[1]The "amount" of a radionuclide (radioactive isotope of an element), especially for those with a "short" physical half-life, is typically given in becquerels (Bq), with 1 Bq being defined as one disintegration (decay) per second. The Bq is an SI-derived unit of activity, with the SI base unit being s^{-1}.

2.3 Brief Characteristics of the Chernobyl Nuclear Accident

The Chernobyl nuclear accident occurred in the early morning hours of April 26, 1986, in the northern Ukrainian SSR, <20 km from the Belarusian border. The Chernobyl nuclear power plant (NPP) was commissioned in 1977 and consisted of four high-power channel-type reactor units (RBMK; Реактор Большой Мощности Канальный) with a power generation capacity of 4 × 1000 MW. The RBMK is a graphite-moderated, water-cooled reactor type that is capable of allowing fuel exchange during operation [thus potentially allowing the "dual-use" of energy production and the production nuclear weapons-grade plutonium-239 (^{239}Pu)]. Two more reactor units had been under construction at the time of the accident, but were never finished after the accident.

The accident happened in the course of a risky experiment in reactor Unit 4 (which had been put into operation only 3 years earlier) that led to the worst civilian nuclear accident in the history of mankind.

In the (ultimately failed) experiment, it was attempted to test whether and how the coasting turbine could be used to provide electricity for the operation of the cooling water pumps of the reactor in case of an electric blackout. The objective of the experiment was to use the coasting turbine of a scrammed reactor as a power source until the emergency diesel generators would provide the necessary power to continue cooling of the reactor. Unsafe operation at low power level led to "xenon poisoning" of the reactor (i.e., the onset of the strong neutron absorber ^{135}Xe within the fuel), which cause the power level to drop. Faulty operation led to oversteering of the reactor. In combination with reactor design flaws (the so-called positive void coefficient, which allows a reactor to increase its power level with increasing temperature) rendered the reactor out of control causing prompt criticality and a massive exceedance of the design power level. The sudden power excursion was equivalent to an estimated 200 tons of TNT [8]. The accident caused several steam and hydrogen explosions and destroyed the reactor building. No containment was in place to prevent the release of radionuclides, as the Soviet position was that such containments were not necessary [9]. The extreme heat of the molten nuclear fuel ignited the graphite moderator of the reactor and caused a fire that lasted for 10 days and was difficult to extinguish. This fire accelerated the release of volatile and less volatile radionuclides into the environment. Enormous efforts and human sacrifices were necessary to control the fire and stabilize the situation, as tons of sand, clay, boron, and lead had been dropped on the disintegrated reactor core [10].

In order to provide a certain degree of stabilization, a concrete structure ("sarcophagus" or "shelter") (see Fig. 2.2) was built around the reactor building. The construction had to be performed under extremely hostile conditions and time pressure and was completed in November 1986 [11]. The sarcophagus was built as a temporary solution, and already in 1993, it was concluded that the shelter would not prevent the release of radioactive dust particles and the leaching of soluble radionuclides as water could still enter the structure, thus making a "permanent solution urgent" [12]. Later, it was concluded that the sarcophagus would have withstood a magnitude 4 earthquake until 1996 [13], but the structure deteriorated further in the following years [11].

Fig. 2.2 The "old" sarcophagus of Unit 4 during heavy snowfall on October 28, 2016, shortly before the New Safe Confinement was moved into position

Fig. 2.3 Unit 4 with the new safe confinement installed

Finally, the construction of a new shelter, the New Safe Confinement (NSC) started in 2010 and moved into position from November 14 to 29, 2016 (Fig. 2.3). The 1.5 billion Euro construction of the NSC was supported financially by the Chernobyl Shelter Fund, an initiative by the G7 governments and contributions by

27 countries. The NSC was constructed adjacent to the destroyed reactor unit and moved into position on rail tracks. With a height of 108 m and a weight of 31,000 tons, the NSC is the largest moveable land-based structure/building ever constructed by man. Its main purpose is to confine the radioactive remains for the upcoming 100 years.

The releases of radionuclides affected inhabited areas in the vicinity of the NPP, above all the city of Pripyat, and also about 200 smaller villages, which all had to be evacuated. The evacuation of the affected population began 3–11 days after the accident. The evacuation of the 44,000 inhabitants of Pripyat was completed on May 6, which was regarded as late by health physicists [14]. In the initial phase, 116,000 people were evacuated from the most contaminated areas. Finally, the evacuation zone encompassed a total of 4300 km^2 [15].

The most affected/contaminated area of the Chernobyl Exclusion Zone is the so-called Red Forest, which is a 10-km^2 area adjacent to the NPP in the western direction. Immediately after the explosion, the wind carried the first cloud with very short-lived fission products to this area, delivering large doses to organisms in the forest. Conifers in this area died from extreme levels of radiation. Their needles turned red, giving the forest its name.

Reactor Unit 1 of Chernobyl NPP was shut down in 1996, Unit 2 in 1991, and Unit 3 in 2000.

Although the environmental situation has largely stabilized during the past decades, the situation may change for the cooling pond of Chernobyl NPP. The cooling pond is a lake with extensions of approx. 11 × 2.5 km (surface area of 22.9 km^2 and a water volume of 59.6×10^6 m^3 during operation). Since the shutdown of the water supply pumps a few years ago, the cooling pond has gradually lost water causing the water level to drop by several meters. The contact of atmospheric oxygen with the now newly exposed sediments of the cooling pond may cause secondary contaminations as nuclear fuel particles that had been conserved in the anoxic conditions of the muddy sediments at the bottom of the cooling pond, may now start to corrode [16]. This could result in a secondary release of bioavailable radionuclides that had been bound inside the particles up to this point.

2.4 Brief Characteristics of the Fukushima Daiichi Nuclear Accident

Fukushima Daiichi NPP consisted of six reactors, three of which (Units 1, 2, and 3) were in operation at the day of the accident (March 11, 2011). All reactors were boiling water reactors (BWR), with power levels ranging from 460 MW$_e$ (Unit 1) to 784 MW$_e$ (Units, 2, 3, 4, and 5) and 1100 MW$_e$ (Unit 6). The reactors were constructed between 1967 and 1979. Two more reactor units were planned for the site, but these plans were canceled after the Fukushima nuclear accident. All reactors used low enriched uranium (LEU) as nuclear fuel, with the exception of Unit 3, which had been operated by a small fraction (6%) of mixed-oxide fuel (MOX) since September 2010 [17].

The Fukushima nuclear accident occurred in the aftermath of the 9.0-magnitude earthquake off the Japanese Pacific coast on March 11, 2011. This earthquake was the worst ever to be recorded in Japanese history, and it caused a devastating tsunami that struck the Japanese east coast. The earthquake triggered automatic shutdown of the reactors, requiring emergency cooling due to station blackout after the shutdown. The height of the tsunami reached up to 40.5 m (depending on the location), and it exceeded the tsunami wall of Fukushima Daiichi NPP, which was only prepared with a 10-m sea wall [18], by four meters. The tsunami caused a flooding of the NPP area and destroyed the emergency diesel generators and rendered the reactors without cooling. The lack of cooling ultimately caused a nuclear meltdown in the three operating reactors. At this high temperature, steam reacted with the zirconium cladding of the nuclear fuel, which caused the onset of large amounts of hydrogen gas. Upon venting operations, which had become necessary to reduce the overpressure inside the pressure vessels, hydrogen gas was released into the reactor buildings where it could mix with oxygen and caused the explosion of Units 1 and 3. Unexpectedly, also Unit 4 exploded although studies showed the fuel of Unit 4 must have been largely intact, as the specific $^{134}Cs/^{137}Cs$ signature of Unit 4 was not found in the environment [19]. It was found that the hydrogen that led to the explosion of the reactor building of Unit 4 had been produced in Unit 3 and was delivered to Unit 4 through backflow during venting [20, 21]. The amounts of hydrogen produced are 890 kg in Unit 1, 460 kg in Unit 2, and 810 kg in Unit 3, respectively [21]. Unit 2 did not suffer a hydrogen explosion because shrapnel produced from the explosion of Unit 1 damaged the venting valve of Unit 2. However, the overpressure that built up in Unit 2 is assumed to have caused structural damage in the lower part of the reactor, and Unit 2 is responsible for major releases and for the distinct contamination strip in northwestern direction from the reactor (approx. 40 km) [22]. The environmental releases were caused mainly by Unit 2 and by the venting operations in Units 1 and 3. For a detailed description of the sequences of the accident, see [23].

Since the containments of the reactor units were still (largely) intact, no sarcophagi were needed for the reactors of Fukushima Daiichi NPP. However, structures were built on the sites of Units 3 and 4, respectively, to facilitate the work-up operations and the recovery of the nuclear fuel (see Fig. 2.4 for the current situation). In contrast to the Chernobyl sarcophagus, which had to be built using the residual structures of the reactor building as support (with unknown stability), much attention was paid in Fukushima that the stability of these newly built structures would not rely on existing structures of the buildings.

Fig. 2.4 The situation of Fukushima Daiichi NPP in 2018

2.5 Comparison of Chernobyl and Fukushima

2.5.1 Radioactive Releases

The amounts of radioactive substances into the environment are clearly distinct between Chernobyl and Fukushima. Table 2.2 compares the release estimates for both accidents.

The comparison of the release data in this table clearly outlines that Fukushima accounted for approximately 10% of the releases of Chernobyl. If one takes into account that about 80% of the atmospheric releases were blown offshore [56–58], the amount Fukushima's releases effectively affecting the Japanese mainland decreases to about 2% of the releases from Chernobyl. In both cases, the majority of the released amounts belonged to the category of the volatile elements (Kr, Xe, I, Cs, Te), but Chernobyl also released significant amounts of less volatile radionuclides (Sr, Ru, lanthanides, actinides, etc.), mainly in the form of hot particles [59].

Table 2.2 Comparison of the atmospheric release estimates of radionuclides for the nuclear accidents at Chernobyl and Fukushima. The bold values represent the most accepted (most cited or most likely) values

Radionuclide	$T_{1/2}$	Activity (PBq)			
		Chernobyl	Reference	Fukushima[a] (atmospheric releases)	Reference
Noble gases					
^{85}Kr	10.75 y	33	[24]	44	[25]
^{133}Xe	5.25 d	6500	[24]	**14,000**	[26]
				15,300	[27]
Volatile elements					
^{3}H	12.3a	1.4 (inventory)	[28]		
^{129m}Te	33.6 d	240	[24]	~15	[3][b]
^{132}Te	3.20 d	**~1150**	[29]	~180	[3][c]
		1000	[24]		
^{129}I	15.7E6 y	**4×10^{-5}–4.8×10^{-5}**	[30, 31]	**5.7×10^{-6}**	[32]
		8.4×10^{-6}	[33]	6.6×10^{-6}	[3][d]
^{131}I	8.03 d	**~1760**	[29]	**150**[e]	[34]
		1200–1700	[24]	130–160	[35]
				190–380	[36]
				65.2	[37]
				200	[38]
				110.7–151	[39]
				105.9	[40]
^{133}I	20.8 h	**910**	[29]	146	[3][f]
		2500	[24]		
^{134}Cs	2.07 y	~47	[24, 29]	**11.8**	[3][g]
				18	[35]
^{136}Cs	13.0 d	36	[24]	**2.6**	[3][h]
				2.2	[3][i]
^{137}Cs	30.1 d	**85**	[29]	**12**[f]	[34, 36]
		74–85	[24]	13	[38]
		98	[41]	6.1–15	[35]
				17	[37]
				36.6	[27]
				35.9	[42]
				15.5	[40]
				8.12	[43]
				19.3	[44]
				9.8–14.5	[39]

(continued)

Table 2.2 (continued)

Radionuclide	$T_{1/2}$	Activity (PBq) Chernobyl	Reference	Fukushima[a] (atmospheric releases)	Reference
Elements with intermediate volatility					
^{89}Sr	50.5 d	**~115**	[29]	~0.2	[3][j]
		81	[24]		
^{90}Sr	28.9 d	**~10**	[29]	~0.02	[3][k]
		4	[31]		
		8	[24]		
^{103}Ru	39.2 d	**>168**	[29]		
		170	[24]		
^{106}Ru	372 d	**>73**	[29]		
		30	[24]		
^{140}Ba	12.8 d	**240**	[29]		
		170	[24]		
Refractory elements[l]					
^{95}Zr	64.0 d	**84**	[29]		
		87	[31]		
		170	[24]		
^{99}Mo	66.0 h	**>72**	[29]		
		210	[24]		
^{125}Sb	2.76 y	0.23	[31]		
^{141}Ce	32.5 d	**84**	[29]		
		200	[24]		
^{144}Ce	285 d	**~50**	[29]		
		140	[24]		
^{154}Eu	8.60 y	0.13	[31]		
^{239}Np	2.36 d	**400**	[29]		
		1700	[24]		
^{238}Pu	87.7 y	**0.015**	[29]	2×10^{-6}-5×10^{-6}	[3][m]
		0.03	[24]		
^{239}Pu	24,100 y	0.013	[29, 31]		
^{240}Pu	6560 y	0.018	[29, 31]		
$^{239+240}Pu$		0.031	[29]	1.0×10^{-6}-2.4×10^{-6}	[45]
^{241}Pu	14.3 y	~2.6	[29]	1.1×10^{-4}-2.6×10^{-4}	[45]
^{242}Pu	3.76E5 y	4×10^{-5}	[29]		
^{241}Am	433 y	0.0024	[31]		
^{242}Cm	163 d	~0.4	[29]		
^{244}Cm	18.1 y	0.0027	[2]		
Total (excluding noble gases)		**~5 300**[e]	[2]	**~520 (430–780)**	[3]

Taken from [3], reprinted with permission from Elsevier 2014

[a]Comment: If necessary, activities of very short-lived radionuclides were decay-corrected to March 12, 2011 (12:00 noon), which is the date of the first releases of radionuclides

Table 2.2 (continued)

[b]Estimated from the ^{137}Cs source term from Chino et al. [34] and the measured $^{129m}Te/^{137}Cs$ activity ratio of 1.3 from Endo et al. [46] (disregarding some obvious outliers). The proposed $^{129m}Te/^{137}Cs$ activity ratio of 4.0 by Tagami et al. [47] has been found to be inconsistent with the other radiotellurium/radiocesium activity ratios tested in this study. This may be due to chemical fractionation between Cs and Te in the environment

[c]Estimated from the ^{137}Cs source term from Chino et al. [34] and the measured $^{132}Te/^{137}Cs$ activity ratio of 15 from Endo et al. [46] (disregarding some obvious outliers)

[d]Based on the ^{131}I data from Chino et al. [34] and the measured atomic ratio for $^{129}I/^{131}I = 31.6$ from Miyake et al. [48]

[e]Most cited value in literature as of May 2013

[f]Based on the ^{131}I data from Chino et al. [34] and a measured $^{133}I/^{131}I$ activity ratio of 0.97 from Amano et al. [49]

[g]Based on the ^{137}Cs data from Chino et al. [34] and a measured $^{134}Cs/^{137}Cs$ activity ratio of 0.98 from Merz et al. [19]

[h]Based on the ^{137}Cs data from Chino et al. [34] and a measured $^{136}Cs/^{137}Cs$ ratio of 0.22 from Tagami et al. [47] as well as Steinhauser et al. [50]

[i]Based on the ^{137}Cs data from Chino et al. [34] and a measured $^{136}Cs/^{137}Cs$ ratio of 0.18 from Amano et al. [49]

[j]Based on the estimation for ^{90}Sr in this study and an initial $^{89}Sr/^{90}Sr$ activity ratio of 11.8 from Povinec et al. [51–53]

[k]Estimated from the ^{137}Cs source term from Chino et al. [34] and the ^{90}Sr–^{137}Cs correlation from Steinhauser et al. [54] (disregarding one outlier)

[l]Chernobyl: releases based on the respective radionuclide inventory in Unit 4 and a release of ~1.5% of the fuel in particulate form [31]

[m]Based on the $^{239+240}Pu$ data from Zheng et al. [45] and a predicted $^{238}Pu/^{239+240}Pu$ activity ratio of 1.92 from Schwantes et al. [55]

2.5.2 *Contaminated Areas*

While Chernobyl's Exclusion zone initially encompassed a 30-km radius (2800 km^2), it was later expanded to 4300 km^2. In contrast, the evacuation zone in Fukushima is largely restricted to the zone of major contamination, i.e., a strip of about 40 km length and 10 km width in the northwestern direction. The areas with a deposition of more than 100 kBq·m^{-2} ^{137}Cs were 56,000 km^2 for Chernobyl and approx. 3000 km^2 in case of Fukushima [60]. Remediation efforts in Fukushima were/are very high—and came/come at a very high cost. They included topsoil removal in contaminated gardens and residential areas and resulted in the opening of formerly evacuated residential areas such as Iitate Village, Minamisōma Town, or Tomioka Town. It is expected that the evacuation zone will be lifted for further settlements in the future. With the exception of the Chernobyl NPP site, comparable efforts had not been undertaken in Chernobyl, and with more than 30 years, the settlements have largely decayed and become uninhabitable. Moreover, the different composition of radionuclide contaminations—in particular the presence of countless highly radioactive and highly dose relevant fuel particles ("hot particles") makes a complete removal of those particles and thus a return to Chernobyl virtually impossible. Since the fuel particles contain long-lived, alpha-emitting actinides and their decomposition rate is partly very slow [61], they will pose a radiation hazard for a long time.

2.5.3 Health Aspects

The health aspects are very complex subject matter and shall be discussed herein only briefly. No doubt remains that Chernobyl exceeded the negative health consequences of Fukushima by far. The most obvious difference is the striking discrepancy concerning the acute (deterministic) radiation effects after both accidents. Chernobyl caused 134 cases of acute radiation syndrome (ARS). In 1986, at least 28 fatalities were due to ARS and the explosions of the reactor [2, 62]. From 1984 to 2004, another 19 workers ("liquidators") died of various causes. In Fukushima, in stark contrast, no acute fatalities were reported, nor any cases of ARS. The stochastic effects, most importantly an increased cancer risk, are much more difficult to assess. Without question, however, (internal) exposure to radioiodine caused an increased thyroid cancer incidence in children and adolescents of more than 7000 additional cases [29]. More recently, an increased leukemia rate was observed among liquidators [63].

In Fukushima, not only the amount of released radioiodine was much lower but also the evacuation and stable iodine prophylaxis [64] worked much more efficiently than in Chernobyl. The United Nations Scientific Committee on the Effects of Atomic Radiation (UNSCEAR) came to the conclusion that an increase in cancer was unlikely observable in Japan after the Fukushima nuclear accident. In particular, the report outlined a low risk of thyroid cancer among children most exposed [65]. Reports of a dramatic increase contradicted this assessment [66]; however, numerous letters to the journal that published this article phrased serious concerns about technical shortcomings of this study. Most importantly, internal exposure through ingestion of contaminated foods can be virtually excluded as a potent dose contributor that would explain an increased cancer risk [67]. The main dose contributor in Chernobyl was ingestion of contaminated food, while the main contributor to the (much smaller) dose in case of Fukushima was external radiation [60, 68]. Several studies have outlined both high levels of contamination in Chernobyl-affected food [69] and a very high degree of food safety in Japan after the Fukushima nuclear accident [35, 70–75] and the implementation of countermeasures such as the usage of potassium fertilizers.

Lastly, the psychological and secondary health effects should not be underestimated for both accidents (see Chaps. 5 and 6 of this book).

2.6 Conclusions

Although both nuclear accidents at Chernobyl and Fukushima were ranked at the maximum level of 7 on the INES, a more detailed analysis shows significant differences between these nuclear emergencies. When comparing factors such as amounts of released radionuclides, contaminated areas, contamination of food, and health effects, it becomes apparent that Chernobyl, by far, exceeded the consequences of the Fukushima nuclear accident.

References

1. McLaughlin TP, Monahan SP, Pruvost NL, Frolov VV, Ryazanov BG, Sviridov VI. A review of criticality accidents. Los Alamos National Laboratory: Los Alamos; 2000.
2. UNSCEAR. Exposures and effects of the Chernobyl accident (annex J). New York: United Nations; 2000.
3. Steinhauser G, Brandl A, Johnson TE. Comparison of the Chernobyl and Fukushima nuclear accidents: a review of the environmental impacts. Sci Total Environ. 2014;470–471:800–17. https://doi.org/10.1016/j.scitotenv.2013.10.029.
4. Jones S. Windscale and Kyshtym: a double anniversary. J Environ Radioact. 2008;99:1–6.
5. Garland JA, Wakeford R. Atmospheric emissions from the Windscale accident of October 1957. Atmos Environ. 2007;41:3904–20. https://doi.org/10.1016/j.atmosenv.2006.12.049.
6. International Atomic Energy Agency. The radiological accident in Goiânia. Vienna: IAEA; 1988.
7. Rogovin M, Frampton GT, Cornell EK, DeYoung RC, Budnitz R, Norry P. Three Mile Island: a report to the commission and the public. NUREG/CR 1250 volume 1, US NRC, Washington, DC; 1980. 184 pp.
8. Malko MV. Recent research activities about the Chernobyl NPP accident in Belarus, Urkraine and Russia. KURRI-KR-79. 2002. pp. 11–27.
9. Marples DR. Chernobyl and nuclear power in the USSR. New York: Palgrave Macmillan; 1986.
10. Gorbachev M. Chernobyl 25 years later: many lessons learned. Bull At Sci. 2011;67:77–80. https://doi.org/10.1177/0096340211399746.
11. Smith JT, Beresford NA. Chernobyl - catastrophe and consequences. Chichester: Springer-Praxis Publishing; 2005.
12. Borovoi A, Gagarinski A. Chernobyl sarcophagus yesterday and today. Nucl Eng Int. 1993;38:21–3.
13. Borovoi A. The sarcophagus: what do we know? What should we do? Nucl Eng Int. 1996;41:28–30.
14. Pröhl G, Mück K, Likhtarev I, Kovgan L, Golikov V. Reconstruction of the ingestion doses received by the population evacuated from the settlements in the 30-km zone around the Chernobyl reactor. Health Phys. 2002;82:173–81. https://doi.org/10.1097/00004032-200202000-00004.
15. IAEA. Facts: the accident was by far the most devastating in the history of nuclear power. 1996. http://www.iaea.org/Publications/Booklets/Chernoten/facts.html. Accessed Sept 2018.
16. Steinhauser G. Anthropogenic radioactive particles in the environment. J Radioanal Nucl Chem. 2018;318:1629–39. https://doi.org/10.1007/s10967-018-6268-4.
17. Nuclear Street. Fukushima to restart using MOX fuel for first time. 2010. http://nuclearstreet.com/nuclear_power_industry_news/b/nuclear_power_news/archive/2010/09/17/fukushima-to-restart-using-mox-fuel-for-first-time-091704#.XIk37KBCcpc. Accessed Jan 2019.
18. Lipscy PY, Kushida KE, Incerti T. The Fukushima disaster and Japan's nuclear plant vulnerability in comparative perspective. Environ Sci Technol. 2013;47:6082–8. https://doi.org/10.1021/es4004813.
19. Merz S, Steinhauser G, Hamada N. Anthropogenic radionuclides in Japanese food: environmental and legal implications. Environ Sci Technol. 2013;47:1248–56. https://doi.org/10.1021/es3037498.
20. INPO. Special report on the nuclear accident at the Fukushima Dai-ichi nuclear power stations. Technical Report 11-005. 2011.
21. Yanez J, Kuznetsov M, Souto-Iglesias A. An analysis of the hydrogen explosion in the Fukushima-Daiichi accident. Int J Hydrog Energy. 2015;40:8261–80. https://doi.org/10.1016/j.ijhydene.2015.03.154.
22. Zheng J, Tagami K, Bu W, Uchida S, Watanabe Y, Kubota Y, et al. Isotopic ratio of $^{135}Cs/^{137}Cs$ as a new tracer of radiocesium released from the Fukushima nuclear accident. Environ Sci Technol. 2014;48:5433–8.

23. Povinec PP, Hirose K, Aoyama M. Fukushima accident: radioactivity impact on the environment. Amsterdam: Elsevier; 2013.
24. Dreicer M, Aarkrog A, Alexakhin R. Et al. consequences of the Chernobyl accident for the natural and human environments. In: IAEA, editor. One decade after Chernobyl: summing up the consequences of the accident. Vienna: IAEA; 1996. p. 319–61.
25. Ahlswede J, Hebel S, Ross JO, Schoetter R, Kalinowski MB. Update and improvement of the global krypton-85 emission inventory. J Environ Radioact. 2013;115:34–42. https://doi.org/10.1016/j.jenvrad.2012.07.006.
26. Stohl A, Seibert P, Wotawa G. The total release of xenon-133 from the Fukushima Dai-ichi nuclear power plant accident. J Environ Radioact. 2012;112:155–9. https://doi.org/10.1016/j.jenvrad.2012.06.001.
27. Stohl A, Seibert P, Wotawa G, Arnold D, Burkhart JF, Eckhardt S, et al. Xenon-133 and caesium-137 releases into the atmosphere from the Fukushima Dai-ichi nuclear power plant: determination of the source term, atmospheric dispersion, and deposition. Atmos Chem Phys Discuss. 2011;11:28319–94. https://doi.org/10.5194/acpd-11-28319-2011.
28. Kirchner G, Noack CC. Core history and nuclide inventory of the Chernobyl core at the time of accident. Nucl Saf. 1988;29:1–5.
29. UNSCEAR. Sources and effects of ionizing radiation (annex D). New York: United Nations; 2008.
30. Aldahan A, Alfimov V, Possnert G. ^{129}I anthropogenic budget: major sources and sinks. Appl Geochem. 2007;22:606–18. https://doi.org/10.1016/j.apgeochem.2006.12.006.
31. Kashparov VA, Lundin SM, Zvarych SI, Yoshchenko VI, Levchuk SE, Khomutinin YV, et al. Territory contamination with the radionuclides representing the fuel component of Chernobyl fallout. Sci Total Environ. 2003;317:105–19. https://doi.org/10.1016/S0048-9697(03)00336-X.
32. Hou X, Povinec PP, Zhang L, Shi K, Biddulph D, Chang C-C, et al. Iodine-129 in seawater offshore Fukushima: distribution, inorganic speciation, sources, and budget. Environ Sci Technol. 2013;47:3091–8. https://doi.org/10.1021/es304460k.
33. Paul M, Fink D, Hollos G, Kaufman A, Kutschera W, Magaritz M. Measurement of iodine-129 concentrations in the environment after the Chernobyl reactor accident. Nucl Instrum Meth B. 1987;B29:341–5. https://doi.org/10.1016/0168-583X(87)90262-X.
34. Chino M, Nakayama H, Nagai H, Terada H, Katata G, Yamazawa H. Preliminary estimation of release amounts of ^{131}I and ^{137}Cs accidentally discharged from the Fukushima Daiichi nuclear power plant into the atmosphere. J Nucl Sci Technol. 2011;48:1129–34. https://doi.org/10.3327/jnst.48.1129.
35. Hamada N, Ogino H. Food safety regulations: what we learned from the Fukushima nuclear accident. J Environ Radioact. 2012;111:83–99. https://doi.org/10.1016/j.jenvrad.2011.08.008.
36. Winiarek V, Bocquet M, Saunier O, Mathieu A. Estimation of errors in the inverse modeling of accidental release of atmospheric pollutant: application to the reconstruction of the cesium-137 and iodine-131 source terms from the Fukushima Daiichi power plant. J Geophys Res D Atmos. 2012;117:D05122/1–16. https://doi.org/10.1029/2011JD016932.
37. Ten Hoeve JE, Jacobson MZ. Worldwide health effects of the Fukushima Daiichi nuclear accident. Energy Environ Sci. 2012;5:8743–57. https://doi.org/10.1039/c2ee22019a.
38. Kobayashi T, Nagai H, Chino M, Kawamura H. Source term estimation of atmospheric release due to the Fukushima Dai-ichi Nuclear Power Plant accident by atmospheric and oceanic dispersion simulations. J Nucl Sci Technol. 2013;50:255–64. https://doi.org/10.1080/00223131.2013.772449.
39. Katata G, Chino M, Kobayashi T, Terada H, Ota M, Nagai H, et al. Detailed source term estimation of the atmospheric release for the Fukushima Daiichi nuclear power station accident by coupling simulations of an atmospheric dispersion model with an improved deposition scheme and oceanic dispersion model. Atmos Chem Phys. 2015;15:1029–70. https://doi.org/10.5194/acp-15-1029-2015.
40. Saunier O, Mathieu A, Didier D, Tombette M, Quélo D, Winiarek V, et al. An inverse modeling method to assess the source term of the Fukushima Nuclear Power Plant accident using gamma

dose rate observations. Atmos Chem Phys. 2013;13:11403–21. https://doi.org/10.5194/acp-13-11403-2013.
41. Anspaugh LR, Catlin RJ, Goldman M. The global impact of the Chernobyl reactor accident. Science. 1988;242:1513–9. https://doi.org/10.1126/science.3201240.
42. Stohl A, Seibert P, Wotawa G, Arnold D, Burkhart JF, Eckhardt S, et al. Xenon-133 and caesium-137 releases into the atmosphere from the Fukushima Dai-ichi nuclear power plant: determination of the source term, atmospheric dispersion, and deposition. Atmos Chem Phys. 2012;12:2313–43. https://doi.org/10.5194/acp-12-2313-2012.
43. Yumimoto K, Morino Y, Ohara T, Oura Y, Ebihara M, Tsuruta H, et al. Inverse modeling of the ^{137}Cs source term of the Fukushima Daiichi nuclear power plant accident constrained by a deposition map monitored by aircraft. J Environ Radioact. 2016;164:1–12. https://doi.org/10.1016/j.jenvrad.2016.06.018.
44. Winiarek V, Bocquet M, Duhanyan N, Roustan Y, Saunier O, Mathieu A. Estimation of the caesium-137 source term from the Fukushima Daiichi nuclear power plant using a consistent joint assimilation of air concentration and deposition observations. Atmos Environ. 2014;82:268–79. https://doi.org/10.1016/j.atmosenv.2013.10.017.
45. Zheng J, Tagami K, Watanabe Y, Uchida S, Aono T, Ishii N, et al. Isotopic evidence of plutonium release into the environment from the Fukushima DNPP accident. Sci Rep. 2012;2:304. https://doi.org/10.1038/srep00304.
46. Endo S, Kimura S, Takatsuji T, Nanasawa K, Imanaka T, Shizuma K. Measurement of soil contamination by radionuclides due to the Fukushima Daiichi nuclear power plant accident and associated estimated cumulative external dose estimation. J Environ Radioact. 2012;111:18–27. https://doi.org/10.1016/j.jenvrad.2011.11.006.
47. Tagami K, Uchida S, Uchihori Y, Ishii N, Kitamura H, Shirakawa Y. Specific activity and activity ratios of radionuclides in soil collected about 20 km from the Fukushima Daiichi nuclear power plant: radionuclide release to the south and southwest. Sci Total Environ. 2011;409:4885–8. https://doi.org/10.1016/j.scitotenv.2011.07.067.
48. Miyake Y, Matsuzaki H, Fujiwara T, Saito T, Yamagata T, Honda M, et al. Isotopic ratio of radioactive iodine (^{129}I/^{131}I) released from Fukushima Daiichi NPP accident. Geochem J. 2012;46:327–33.
49. Amano H, Akiyama M, Chunlei B, Kawamura T, Kishimoto T, Kuroda T, et al. Radiation measurements in the Chiba metropolitan area and radiological aspects of fallout from the Fukushima Dai-ichi nuclear power plants accident. J Environ Radioact. 2012;111:42–52. https://doi.org/10.1016/j.jenvrad.2011.10.019.
50. Steinhauser G, Merz S, Hainz D, Sterba JH. Artificial radioactivity in environmental media (air, rainwater, soil, vegetation) in Austria after the Fukushima nuclear accident. Environ Sci Pollut Res. 2013;20:2527–37. https://doi.org/10.1007/s11356-012-1140-5.
51. Casacuberta N, Masqué P, Garcia-Orellana J, Garcia-Tenorio R, Buesseler KO. ^{90}Sr and ^{89}Sr in seawater off Japan as a consequence of the Fukushima Dai-ichi nuclear accident. Biogeosci Discuss. 2013;10:2039–67.
52. Nishihara K, Yamagishi I, Yasuda K, Ishimori K, Tanaka K, Kuno T, et al. Radionuclide release to stagnant water in Fukushima-1 nuclear power plant. Atmos Energy Soc Jpn. 2012;11:13–9.
53. Povinec PP, Hirose K, Aoyama M. Radiostrontium in the Western North Pacific: characteristics, behavior, and the Fukushima impact. Environ Sci Technol. 2012;46:10356–63. https://doi.org/10.1021/es301997c.
54. Steinhauser G, Schauer V, Shozugawa K. Concentration of strontium-90 at selected hot spots in Japan. PLoS One. 2013;8:e57760. https://doi.org/10.1371/journal.pone.0057760.
55. Schwantes JM, Orton CR, Clark RA. Analysis of a nuclear accident: fission and activation product releases from the Fukushima Daiichi nuclear facility as remote indicators of source identification, extent of release, and state of damaged spent nuclear fuel. Environ Sci Technol. 2012;46:8621–7. https://doi.org/10.1021/es300556m.
56. Aoyama M. Long-range transport of radiocaesium derived from global fallout and the Fukushima accident in the Pacific Ocean since 1953 through 2017—part I: source term

and surface transport. J Radioanal Nucl Chem. 2018;318:1519–42. https://doi.org/10.1007/s10967-018-6244-z.
57. Kawamura H, Kobayashi T, Furuno A, In T, Ishikawa Y, Nakayama T, et al. Preliminary numerical experiments on oceanic dispersion of ^{131}I and ^{137}Cs discharged into the ocean because of the Fukushima Daiichi nuclear power plant disaster. J Nucl Sci Technol. 2011;48:1349–56. https://doi.org/10.1080/18811248.2011.9711826.
58. Morino Y, Ohara T, Nishisawa M. Atmospheric behavior, deposition, and budget of radioactive materials from the Fukushima Daiichi nuclear power plant in March 2011. Geophys Res Lett. 2011;38:L00G11/1–7. https://doi.org/10.1029/2011GL048689.
59. Beresford NA, Fesenko S, Konoplev A, Skuterud L, Smith JT, Voigt G. Thirty years after the Chernobyl accident: what lessons have we learnt? J Environ Radioact. 2016;157:77–89. https://doi.org/10.1016/j.jenvrad.2016.02.003.
60. Howard BJ, Fesenko S, Balonov MI, Pröhl G, Nakayama S. A comparison of remediation after the Chernobyl and Fukushima Daiichi accidents. Radiat Prot Dosim. 2017;173:170–6. https://doi.org/10.1093/rpd/ncw312.
61. Kashparov V, Yoschenko V, Levchuk S, Bugai D, Van Meir N, Simonucci C, et al. Radionuclide migration in the experimental polygon of the Red Forest waste site in the Chernobyl zone – part 1: characterization of the waste trench, fuel particle transformation processes in soils, biogenic fluxes and effects on biota. Appl Geochem. 2012;27:1348–58. https://doi.org/10.1016/j.apgeochem.2011.11.004.
62. The Chernobyl Forum. Chernobyl's legacy: health, environmental and socio-economic impacts and recommendations to the governments of Belarus, the Russian Federation and Ukraine. Vienna: IAEA; 2006.
63. Cardis E, Hatch M. The Chernobyl accident - an epidemiological perspective. Clin Oncol. 2011;23:251–60.
64. Nishikawa Y, Kohno A, Takahashi Y, Suzuki C, Kinoshita H, Nakayama T, et al. Stable iodine distribution among children after the 2011 Fukushima nuclear disaster in Japan: an observational study. J Clin Endocrinol Metab. 2019;104(5):1658–66. https://doi.org/10.1210/jc.2018-02136.
65. United Nations Scientific Committee on the Effects of Atomic Radiation (UNSCEAR). UNSCEAR 2013 report, volume I. Report to the general assembly, scientific annex A: levels and effects of radiation exposure due to the nuclear accident after the 2011 great east-Japan earthquake and tsunami. 2013. www.unscear.org/docs/reports/2013/13-85418_Report_2013_Annex_A.pdf.
66. Tsuda T, Tokinobu A, Yamamoto E, Suzuki E. Thyroid cancer detection by ultrasound among residents ages 18 years and younger in Fukushima, Japan: 2011 to 2014. Epidemiology. 2016;27:316–22. https://doi.org/10.1097/EDE.0000000000000385.
67. Steinhauser G, Chávez-Ortega M, Vahlbruch J-W. Japanese food data challenge the claimed link between Fukushima's releases and recently observed thyroid cancer increase in Japan. Sci Rep. 2017;7:10722. https://doi.org/10.1038/s41598-017-10584-8.
68. Harada KH, Niisoe T, Imanaka M, Takahashi T, Amako K, Fujii Y, et al. Radiation dose rates now and in the future for residents neighboring restricted areas of the Fukushima Daiichi nuclear power plant. Proc Natl Acad Sci U S A. 2014;111:E914–23. https://doi.org/10.1073/pnas.1315684111.
69. Travnikova IG, Bruk GJ, Shutov VN, Bazjukin AB, Balonov MI, Rahola T, et al. Contribution of different foodstuffs to the internal exposure of rural inhabitants in Russia after the Chernobyl accident. Radiat Prot Dosim. 2001;93:331–9. https://doi.org/10.1093/oxfordjournals.rpd.a006445.
70. Hamada N, Ogino H, Fujimichi Y. Safety regulations of food and water implemented in the first year following the Fukushima nuclear accident. J Radiat Res. 2012;53:641–71.
71. Harada KH, Fujii Y, Adachi A, Tsukidate A, Asai F, Koizumi A. Dietary intake of radiocesium in adult residents in Fukushima prefecture and neighboring regions after the Fukushima nuclear power plant accident: 24-h food-duplicate survey in December 2011. Environ Sci Technol. 2012;47:2520–6. https://doi.org/10.1021/es304128t.

72. Koizumi A, Harada KH, Niisoe T, Adachi A, Fujii Y, Hitomi T, et al. Preliminary assessment of ecological exposure of adult residents in Fukushima prefecture to radioactive cesium through ingestion and inhalation. Environ Health Prev Med. 2012;17:292–8.
73. Merz S, Shozugawa K, Steinhauser G. Analysis of Japanese radionuclide monitoring data of food before and after the Fukushima nuclear accident. Environ Sci Technol. 2015;49:2875–85. https://doi.org/10.1021/es5057648.
74. Steinhauser G. Assessment of the effectiveness of the post-Fukushima food monitoring campaign in the first year after the nuclear accident: a hypothesis. J Environ Radioact. 2016;151:136–43. https://doi.org/10.1016/j.jenvrad.2015.09.013.
75. Steinhauser G, Saey PRJ. Cesium-137 in the meat of wild boars: a comparison of the impacts of Chernobyl and Fukushima. J Radioanal Nucl Chem. 2016;307:1801–6. https://doi.org/10.1007/s10967-015-4417-6.

Chapter 3
Lessons Learned from the Chernobyl Accident

Valery Kashparov

Abstract This chapter describes the lessons learned from the Chernobyl accident on the basis of practical experience. Main characteristics of radionuclides release and consequences of radiological contamination of the environment, and also remediation actions, taken to protect workers and population against radiation at different stages of rectification of the consequences of the accident in Belarus, Russia, and Ukraine in 1986–2018 are analyzed. Criteria for applying countermeasures, such as maximum expected effective irradiation dose for the population and terrestrial density of radionuclides contamination for evacuation and resettlement, restriction of business activities, etc., and also action level of radionuclides in food to reduce the internal dose, are provided. Main positive and negative features of the decisions taken in the process of the Chernobyl nuclear disaster elimination are considered. Practically all agricultural countermeasures implemented in the large scale on contaminated lands after Chernobyl accident can be recommended for use in case of future accidents. We focus mainly on the Chernobyl exclusion zone as the territory of radiation-ecological reserves of Ukraine and Belarus for scientific research in the field of radioecology and radiobiology, as well as on the most contaminated 10-km zone around the Chernobyl nuclear power plant—a Zone for special industrial usage, not suitable for living in the near future.

By an example of the Chernobyl disaster, it is shown that in comparison with radiological consequences the socio-psychological ones have made much more influence on human life and health due to lack of urgent, objective, and truthful information on the accident and its impact on the health, in mass media.

Keywords Radionuclides · Radioecology · Radiobiology · Remedial action · Milk contamination · The Chernobyl accident · Effective dose · The Chernobyl exclusion zone

---

V. Kashparov (✉)
Ukrainian Institute of Agricultural Radiology, National University of Life and Environmental Sciences of Ukraine, Kyiv, Ukraine

CERAD CoE Environmental Radioactivity/Department of Environmental Sciences, Norwegian University of Life Sciences, Aas, Norway

G. Steinhauser et al. (eds.), *Nuclear Emergencies*,
Current Topics in Environmental Health and Preventive Medicine,
https://doi.org/10.1007/978-981-13-8327-4_3

3.1 Introduction

As a result of sharp increase of neutron flux with a subsequent release of energy on April 26, 1986, at 01.24 am, the 4th block of Chernobyl NPP (ChNPP) was destroyed. Gaseous (Kr and Xe radioisotopes) and volatile (I, Ag, Cs and Te) fission products, as well as particles of irradiated nuclear fuel, containing non-volatile radionuclides (Sr, Ru, Zr, Nb, Sb, Ba, Ce, Eu, Pu, Am, etc.), were released into the atmosphere [1–4]. Due to the high temperature of nuclear fuel and construction materials, caused by the release of energy at nuclear decay, the oxidation and destruction of UO_2 occurred up to May 5, 1986, and as a result, leakage and high rise of volatile fission products and fuel particles (FP) in the convective plume were observed. The most relevant fission and activation products and their half-lives ($T_{1/2}$) are: ^{90}Sr ($T_{1/2}$ = 29 y), ^{131}I ($T_{1/2}$ = 8.0 d), ^{134}Cs ($T_{1/2}$ = 2.1 y) ^{137}Cs ($T_{1/2}$ = 30.2 y), ^{238}Pu ($T_{1/2}$ = 87.7 y), ^{239}Pu ($T_{1/2}$ = 24,100 y), ^{240}Pu ($T_{1/2}$ = 6563 y), ^{241}Pu ($T_{1/2}$ = 14.3 y), and ^{241}Am ($T_{1/2}$ = 432.8 y), respectively. Over the half of iodine radioisotopes (~1760 PBq of ^{131}I), one-third of cesium radioisotopes (~85 PBq of ^{137}Cs) and under 2% of non-volatile long-lived radionuclides within fuel particles (~4 PBq of ^{90}Sr, 0.046 PBq of $^{238\text{-}240}$Pu, 0.0024 PBq of ^{241}Am, and 2.6 PBq of ^{241}Pu) were released from reactor during the Chernobyl accident on April 26—May 5, 1986 [4–6]. As a result, over 200,000 km^2 of European territory (~65,000 km^2 in Russia, ~46,000 km^2 in Belarus, ~43,000 km^2 in Ukraine, ~23,000 km^2 in Sweden, ~19,000 km^2 in Finland, ~11,000 km^2 in Austria, ~7000 km^2 in Norway, etc.) were contaminated with the long-lived ^{137}Cs above 40 kBq m^{-2} [1], with the highest deposition levels found in the Chernobyl exclusion zone (ChEZ)—Fig. 3.1 [4, 7].

Fuel particles with irradiated uranium oxide matrix with various impurities—one of key features of Chernobyl nuclear fallout—were observed not only close to ChNPP but also at a considerable distance—in various European countries [2, 7–10]. Due to the high speed of dry gravitational deposition of the fuel particles (density 8–10 g cm^{-3}) in the atmosphere, mostly ChEZ and adjacent territories (Fig. 3.1) were contaminated with the radionuclides of fuel components of Chernobyl radioactive fallout (^{90}Sr, $^{238\text{-}241}$Pu, ^{241}Am, etc.). Before the Chernobyl accident, there was no information concerning the behavior of radionuclides released within fuel particles into the environment [7, 8, 10–13]. The use of migration parameters for water-soluble radiostrontium forms obtained after the Kyshtym accident in the Southern Urals in 1957, as well as in the laboratory experiments and after global radioactive fallout, led to very conservative estimates of surface water and vegetation contamination. Based on these assessments, protective measures were promptly implemented (construction of dams, bank balancing, etc.); however, their effectiveness was extremely low and there was no real need for it [1].

Most of all, the Chernobyl disaster affected the rural population and agricultural production in Belarus, Russia, and Ukraine. Radioactive contamination of agricultural lands, semi-natural pastures, and hayfields led to radioactive pollution of food products and increased doses of internal irradiation of the rural population.

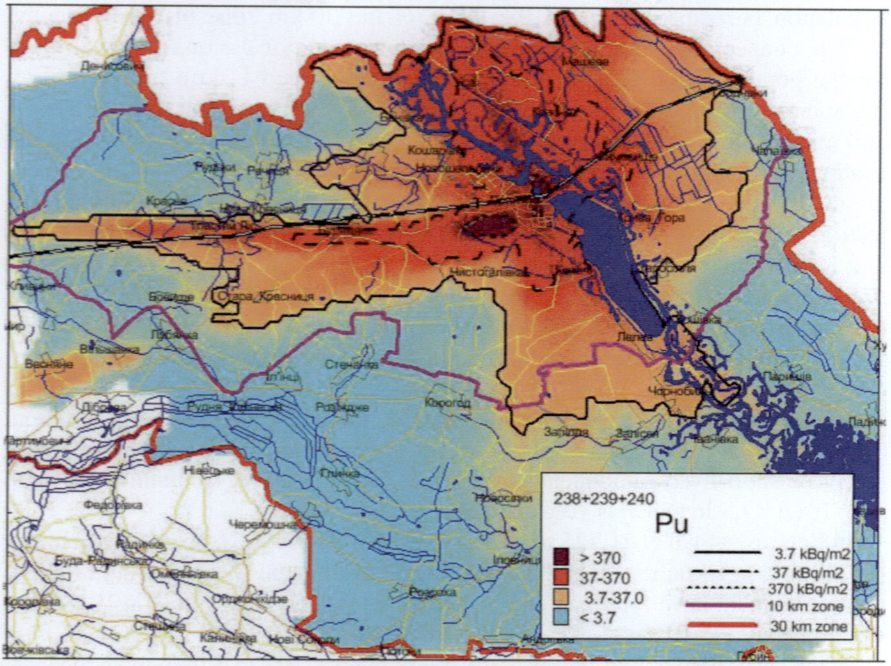

Fig. 3.1 Terrestrial density of $^{238+239+240}$Pu contamination of the Chernobyl Exclusion Zone in 2018 (developed on the basis of [7]). © by the author

3.2 The Early Phase of Chernobyl Accident

The highest doses were received by the personnel (firefighters, station employees, doctors, etc.), who were directly involved in rectification of the consequences during the first hours and days of the ChNPP accident. The diagnosis 'acute radiation syndrome' (ARS) was confirmed in 134 people (the absorbed dose was 0.8–16 Gy), among them 28 died in 1986 (out of 21 people with the highest doses of 6.5–16 Gy with the diagnosis very severe (IV) ARS 20 people died). Nineteen people with ARS diagnosed in 1986 died in 1987–2004 from causes not (directly) related to radiation. The extremely severe consequences of personnel exposure were mostly caused by the lack of personal protective equipment for the skin from beta radiation and respiratory organs from radioactive aerosols (jumpsuits, respirators, etc.).

From April 26 until May 6, 1986, when the main fallout took place and radioactive contamination of the territory was formed, in order to prevent severe radiation damage, the evacuation of the population (99,195 people) from 113 settlements (51 in Belarus and 62 in Ukraine) km zone around the Chernobyl NPP including the city of Pripyat was organized. Stable iodine prophylaxis was not done at the right time. The analysis of the radiation situation after the radioactive emissions showed that the territory of radioactive contamination, where radiation protection of the population is required, exceeds the limits of the 30-km zone of the Chernobyl NPP. In connection with this, in summer 1986 the additional evacuation was made for the population of villages where the expected effective dose from April 26, 1986, until April 25, 1987, could exceed the established temporary annual limit—100 mSv (50 mSv from external and 50 mSv from internal exposure). For external irradiation zonation, the dose rate limit was used on May 10, 1986–5 mR h^{-1} (about 50 μSv h^{-1}). To avoid the exceedance of the internal irradiation dose limit, the terrestrial density of contamination of 555 kBq m^{-2} with ^{137}Cs, 111 kBq m^{-2} with ^{90}Sr, and 3.7 kBq m^{-2} with $^{239,240}Pu$ was used as a criterion. In summer 1986, an additional evacuation of the population was provided from 51 settlements in Belarus and 15 settlements in Ukraine, where the dose rate exceeded 5 mR h^{-1}. The ratio between the contamination of the territory with short-lived gamma-emitting radionuclides (^{95}Zr, ^{95}Nb, $^{103,106}Ru$, $^{141,144}Ce$), released as a part of FP, and the long-lived condensation component of ^{137}Cs at different points of the exclusion zone was different. Therefore, with the same dose rate—5 mR h^{-1} in 1986, at the present time, after the decay of short-lived radionuclides, the terrestrial density of contamination with long-lived ^{137}Cs in ChEZ is very different.

In Belarus, a total of 24,725 people were evacuated in 1986 from 108 settlements (1542 km^2); in Ukraine—91,406 people from 75 settlements (2157 km^2). In Russia, the resettlement from four locations (186 people) was organized only in 1988 from the area of 193 km^2.

In the acute period after the accident, it was not possible to differentiate the level of contamination in animals and during the period of May–July 1986, the total number of slaughtered animals reached 95,500 cattle and 23,000 pigs. Many carcasses were buried and some were stored in refrigerators, which created great hygienic, practical, and economical difficulties. A technique for in vivo measurements of ^{137}Cs in animals (live monitoring of animals) with application of the clean fodder could reduce the radionuclides activities in meat was developed and used since 1987 [1, 14, 15]. In absence of this method, in face of a lack of clean forage for the evacuated animals and difficulties in managing large numbers of animals, and to prevent the psychological influence to population in the case of possible death of animals, more than 100,000 of agricultural animals were slaughtered.

In late April/early May 1986 in Belarus and Ukraine, dairy cows were already grazing outdoors, and there were significant levels of radionuclides activity concentration in cow milk. At the early phase, ^{131}I (half-life is 8 days) was the main contributor to the population internal dose through the pasture-cow-milk pathway. Peak concentrations occurred rapidly (within about 1 day) after deposition (in late April or early May 1986, depending on when deposition happened in certain places). In Ukraine, activity concentrations of ^{131}I in milk exceeded action levels (3700 Bq L^{-1} from May 6, 1986—[16]), which ranged from a few hundred to a few ten thousand Becquerel per liter. The activity concentration of ^{131}I in milk decreased with an effective half-life of 4–7 days owing to its short physical half-life and the processes that removed it from leaves [17]. Consumption of leafy vegetables onto which radionuclides had been deposited also contributed to the intake of radionuclides by humans. Radiation monitoring of the agricultural production contamination was arranged in 1–2 weeks after the beginning of the accident at the large milk plants and in the collective farms. Urban population was mainly protected against consumption of the radioactive-contaminated agricultural products, especially milk, through the distribution network (foodstuffs were delivered from the clean regions). Rural populations that had cows in private farms had not been informed about contamination of milk with ^{131}I, which resulted in the high doses to thyroid gland and increase of the thyroid cancer morbidity in children in Belarus and Ukraine after the accident [1, 18]. The information on countermeasures for milk was confined to managers and local authorities and was not distributed to the private farming system of the rural population. This resulted in limited application of the countermeasures with some delay, especially in rural settlements for privately produced milk, resulting in a low effectiveness in some areas.

The main mistake in the acute period of liquidation (i.e., implementation of countermeasures) of the Chernobyl accident was the lack of timely and objective information concerning the urgent radiological situation, the risk to the public health and the need for protective measures for the population and local authorities. Such a "secrecy" regime led to distrust of information from the official mass media and subsequent dissemination of unprofessional and unreliable information.

3.3 Later Phase of Chernobyl Accident

After 5 years since the Chernobyl accident, in early 1991, before the collapse of the USSR, the laws on legal regime of the territories affected by the Chernobyl accident, according to which the zones of radioactive contamination were determined and additional resettlement of the population in 1991–1994 was organized, were adopted in Belarus, Russia, and Ukraine [19, 20]. The main criterion for safe living on the territory contaminated after the Chernobyl disaster was the limit of the average annual effective dose of exposure to the population—1 mSv y^{-1}. Compulsory resettlement was carried out at an average annual effective dose of radiation of above 5 mSv y^{-1} or at a terrestrial density of contamination with radioisotopes of cesium of above 555 kBq m^{-2} or $^{90}Sr > 111$ kBq m^{-2} or $Pu > 3.7$ kBq m^{-2}.

According to the Law of Ukraine, the radiation-hazardous lands include territories where the density of contamination with plutonium isotopes is >3.7 kBq m^{-2}. Living on these territories is prohibited [20]. Probabilistic analysis of the passage of the radiation-hazardous boundaries of the land shows that even after 500–1000 years, the density of $^{238-240}Pu$ contamination will exceed 3.7 kBq m^{-2} in the 10-km zone around the ChNPP (about 450 km^2), and it will not be suitable for living in the foreseeable future—see Fig. 3.1 [7, 19]. In this regard, the possibility of using this territory as the special industrial use area for the radioactive waste management, etc., which will not be subject to regulatory and legal acts for territories with possible population residence, is nowadays being considered.

The increase in radioactive contamination of the environment ^{241}Am in the coming decades due to the radioactive decay of ^{241}Pu will not provide a significant effect on the change in the radiological situation due to the insignificance of this increase (<20%) and simultaneous decrease in the activity of ^{238}Pu.

Thus, 340,000 people were evacuated or relocated in 1986–1994 years from the most contaminated territories of Belarus—6200 km^2, Russia—193 km^2, and Ukraine—4.200 km^2, including 2000 km^2 outside ChEZ of Ukraine, where traditional economic activity was discontinued or mostly restricted [1, 19].

The citizens had the willful right to resettle if the effective radiation dose was over 1 mSv y^{-1} or the terrestrial density of contamination with radioisotopes of cesium was over 185 kBq m^{-2} or ^{90}Sr > 5.55 kBq m^{-2} or Pu > 0.37 kBq m^{-2}. However, due to the absence of reliable data about the doses of exposure in the early 1990s, the main criterion for zonation of the territory was the density of its contamination with $^{134,137}Cs$. Whereas the doses of external exposure to the population in settlements of different types correlated with the density of contamination of the territory with $^{134,137}Cs$, the internal radiation dose, caused by radioactive contamination of food products, depended mostly on the biological mobility of radionuclides due to soil and climatic conditions. Thus, on the waterlogged peat lands of the Rivne region, Ukraine, the abnormally high radiocesium transfer rates into the grass and then into the milk were observed [21, 22]. With a density of contamination with terrestrial ^{137}Cs of about 40 kBq m^{-2}, the average annual exposure dose to the population even after 25–30 years since the disaster could exceed 1 mSv y^{-1} [21, 23].

With the improvement of the radiological situation, the permissible levels of radionuclide content in food and drinking water decreased, and finally reached the non-emergency level in Ukraine in 1997, in Belarus in 1999, and in Russia in 2001 [16].

Nowadays 7818 settlements (2402 in Belarus, 4413 in Russia, and 1003 in Ukraine) are classified as radioactive contamination zones with more than three million residents [19, 24]. According to the dosimetric passportization data, in 2011–2012 in 25–26 settlements in Ukraine, the average annual effective dose of exposure to the population exceeded 1 mSv, and therefore only these populated areas can be considered contaminated, and protective measures should be taken there to reduce the exposure to radiation [18, 20, 23]. In these settlements in the northern part of Ukraine at a distance of 200–300 km to the west of the ChNPP, the average annual effective dose of exposure to the population is above 1 mSv, 70–90% due to internal exposure caused by the consumption of local milk containing ^{137}Cs above the permissible level (PL = 100 Bq L^{-1})—Fig. 3.2 [21, 22, 26]. Due to the absence of governmental programs for the implementation of necessary protective measures to reduce radiological risks to impacted populations, the exceedance of PL for the activity concentration of ^{137}Cs in cow's milk for adults of 100 Bq L^{-1}in the Chernobyl-affected areas of Ukraine could persist for many more years—until at least 2040 [21].

The optimization of the remedial strategy for settlement actions (application of Ferrocyn (hexacyanoferrates) to cows, mineral fertilization of potato fields, information campaigns on consumption of wild mushrooms and other forestry products, and feeding pigs with uncontaminated fodder) exposed to an effective dose above 1 mSv year^{-1} has shown that a diversity of measures can decrease effective dose for a representative person to below 1 mSv year^{-1} and ^{137}Cs activity concentration in

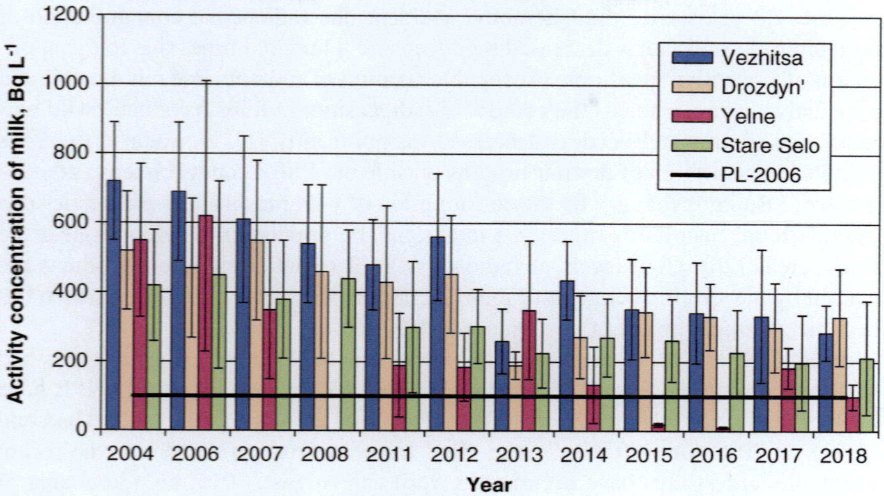

Fig. 3.2 The dynamics of the milk contamination by ^{137}Cs which is produced in the private farms of the most critical settlements of Ukraine during the grazing period (arithmetic mean, standard deviation, $n > 20$) and permissible level (PL-2006) for milk (100 Bq L^{-1}) (developed on the basis of data from [25]). © by the author

milk <100 Bq L^{-1} in Ukraine with moderate financial expenses [21, 27–29]. Unfortunately, since 2009 there have been no protective measures aimed at radiation protection of the population of Ukraine [30].

Due to dissolving the fuel particles and leaching of ^{90}Sr, an increase in its bioavailability was observed during the post-accident period, which has now reached its maximum [8, 13]. Therefore, ^{90}Sr content in food grains and fuel wood in regions close to ChEZ may exceed the established permissible levels [10, 19].

The inhalation intake of radionuclides, as well as of secondary contamination of the territory due to resuspension of radionuclides in the internal exposure, even during agricultural work, was insignificant already after 1 year since the accident (<1%) [31, 32].

Over more than 30 years after the Chernobyl accident, the activity of ^{90}Sr and ^{137}Cs has decreased by more than two times, which led to the reduction of the total area with ^{137}Cs contamination density above 37 kBq m^{-2} in Belarus, Russia, and Ukraine by factors of 1.5, 2.9, and 2.7, respectively [19]. The radiation dose for the population caused by both external and internal exposure decreased significantly due to the reduction of food contamination [18]. Therefore, it is necessary to revise the radioactive zonation of the settlements and to return territories excluded after the Chernobyl accident into the economic use. Whereas in Belarus such a revision of zonation is carried out on the regular basis, in Ukraine, since 1991 there has been no revision on referring the territories to zones of radioactive contamination. Out of more than 900 previously inhabited settlements in the radioactive contamination zone, <5% now can be considered contaminated according to the law [20, 23].

Since 30 years after the Chernobyl accident, the radioactive contamination of agricultural products has decreased by dozen and a hundred times due to the application of protective measures, irreversible fixation of radionuclides in the soil and radioactive decay, whereas the content of radiocesium in forest products (wild animals, mushrooms and berries) decreased insignificantly [1]. Nowadays, the ^{137}Cs activity concentration of dry mushrooms outside the ChEZ can reach tens and hundreds of kBq kg^{-1}, that is why the consumption of wild mushrooms and berries can provide more than half of the ^{137}Cs intake for the population of villages near the forest areas [26]. High levels of radioactive contamination of forest products are caused by the specific behavior of radionuclides in the forest ecosystem, which has become an urgent object of study recently [33–36].

From 1993 to 2018, more than one thousand wildfires of different types and scales, including those in the most contaminated 10-km area near the ChNPP, have been officially recorded in ChEZ: in July 15–17 and 26 July, 2016 (on ~300 ha), and June 5–7, 2018 (on ~50 ha), on the "Red forest" territory [37, 38]. In the recent years, the largest fires have occurred in April and August, 2015, on a total area of about 15,000 hectares of meadows and forest lands [39–41]. The results of active experiments and mathematical modeling show that fires in ChEZ do not make a significant contribution to secondary contamination of territories outside ChEZ and to additional doses of radiation to the population [42–44]. In addition to the radiological hazard of fires in the contaminated area for the firefighters, ChEZ personnel and the population, which is often exaggerated in the media [40, 41], the information about the fires in ChEZ is of great social and psychological importance for the population of Ukraine as well as beyond its borders. In this regard, special attention should be paid to fire prevention measures in ChEZ, as well as to the creation of a modern fire detection and control system [19].

Nowadays the main radiological hazard in ChEZ consists in medium and long-lived radionuclides: ^{90}Sr, ^{137}Cs, ^{238}Pu, ^{239}Pu, ^{240}Pu, and ^{241}Am. In order to forecast the release of radionuclides from ChEZ by water, a routine scientific monitoring of ^{90}Sr and plutonium radioisotopes migration with groundwater from radioactive waste storage is conducted. The obtained results showed that the migration rate of radionuclides with groundwater does not exceed 1 m $year^{-1}$ and does not pose a hazard for radioactive contamination of the main water artery of Ukraine—the Dnieper River [45, 46].

According to the Decree of the president of Ukraine No.. 174/2016 on the territory of ChEZ outside the most contaminated "10-km" zone of special industrial use, in 2016 on the area of 226,964.7 ha, the Chernobyl Radiation and Ecological Biosphere Reserve was established. In Belarus, a similar reserve in ChEZ was also established in 1997 as a buffer zone for radiation protection of the population in the neighboring territories.

The IAEA Chernobyl Forum recommends using ChEZ territory for scientific research in the field of radioecology and radiobiology [1, 45, 47, 48]. The high gradients of radionuclide contamination of the territory, various forms of radioactive fallout, types of soils and landscapes, and large biodiversity make ChEZ a unique ground for studying radiobiological effects for the purpose of radiation protection of the environment [49–56].

The principal mistake in the late phase of liquidation of the Chernobyl accident was compensation-oriented populism in the social protection of the population on the contaminated territories and participants in the liquidation of the consequences of the accident (various cash payments, wage supplements, free meals, etc.) and benefits (early retirement, free allocation of housing, resort and sanatorium rehabilitation, privileges for admission to higher education institutions, etc.) for the potential (not yet realized) risk of exposure to ionizing radiation. The above mentioned, in some cases, did not stimulate the use of protective measures for the purpose of radiation protection of the population.

The use of terrestrial density of radiocesium contamination without taking into account its bioavailability as well as extremely conservative predictive estimates for radioisotopes of strontium and plutonium as auxiliary criteria for radioactive zonation of the territories of Belarus, Russia, and Ukraine led to a discrepancy between the expected and the actual effective exposure doses to the population.

The resettlement from villages in 5–8 years after the accident appeared to be not effective in the radiological and economic terms, when more than 70% of the effective dose had already been received by the population. Also, significant financial resources had already been invested during the post-accident period into the infrastructure for the rehabilitation of these villages [18]).

In our opinion, due to objective reasons, during the entire post-accident period, ChEZ was not sufficiently used for the scientific research. Some potentially important radionuclides, such as ^{36}Cl, ^{99}Tc, ^{129}I remain almost untouched [48, 57–62].

3.4 Lessons

As a result of the Chernobyl accident, about 5300 PBq of radionuclides (excluding noble gases), the most radiologically important—^{131}I and ^{137}Cs, were released into the atmosphere; more than 200,000 km^2 of Europe contaminated with ^{137}Cs, mainly in the USSR; irradiated more than 600,000 "liquidators" [63]; 340,000 people were evacuated or relocated in the years 1986–1994; more than five million people permanently live in the contaminated areas; economic losses in the hundreds of billions of US dollars.

Immediate evacuation of 116 317 people from 187 settlements of exclusion zone (average cumulative effective dose was 33 mSv) was organized promptly and timely.

Application of agricultural countermeasures allowed to decrease twice the effective internal dose to the population. Practically all agricultural countermeasures implemented in the large scale on contaminated lands after the Chernobyl accident can be recommended for use in case of further accidents. However, the effectiveness of most soil-based countermeasures varies at each site. Therefore, analysis of soil properties and agricultural practice before their application is of great importance [1].

About 100,000 of agricultural animals from the 30-km exclusion zone were slaughtered, which created great hygienic, practical, and economic difficulties. The usage of technique for in vivo measurements of ^{137}Cs in animals (live monitoring of animals) with application of the clean fodder could have reduced the contamination of animals to the permissible level in 1986.

At the late stage of the liquidation of the Chernobyl accident, due to the current situation, populism and political interests were often placed above the radiological, economic, and social factors, which led to the prevalence of social protection over radiation protection for the Chernobyl-affected population. This resulted in the dependence of benefits and compensations over the amount of the potential exposure risk (the higher is the risk, the higher are the benefits and compensation); resettlement of the population in 5–8 years after the accident, when more than 70% of the exposure dose had already been received by the population was clearly ineffective (for example, the resettlement of the regional center of the city of Polesskoe in 1994), etc. Excessively conservative criteria for zonation of radioactive contamination of territories according to the effective dose of radiation (0.5 mSv y^{-1} in Ukraine) and radionuclide contamination density (>37 kBq m^{-2} for ^{137}Cs; >5.5 kBq m^{-2} for ^{90}Sr; >0.37 kBq m^{-2} for $^{238-240}Pu$) were used [19]. Despite the objective change in the radiological situation after the Chernobyl accident, the complexity of the regulatory procedure still do not allow to revise the zonation of the contaminated territories of Ukraine.

The most severe consequence for the health of the population after the Chernobyl disaster is an increase in the number of thyroid cancer among children associated with the use of contaminated ^{131}I milk during April–May 1986 (the rural population was not informed about it because of the "secrecy" regime in the USSR). The lack of any information on real and potential radioactive contamination of the environment, the health risks, available protective measures, including restrictions on the consumption of local food products, no dialog with the population were the principle mistakes made after the Chernobyl accident that caused further lack of trust in any official information. For this reason, the socio-psychological consequences of the Chernobyl accident turned out to be more severe than the radiological ones. Up to the present day, the population is much concerned about the radioactive contamination of food in certain regions of Ukraine [21], forest fires in ChEZ [41], etc. This requires constant presentation of urgent and objective information on the existing radiological hazard, as well as its explanation at the national level in accordance with the international standards.

References

1. IAEA. Environmental consequences of the Chernobyl accident and their remediation: twenty years of experience. Report of the Chernobyl forum expert group 'environment'. Vienna: IAEA; 2006.
2. Kashparov VA, Ivanov YA, Zvarich SI, Protsak VP, Khomutinin YV, Kurepin AD, Pazukhin EM. Formation of hot particles during the Chernobyl nuclear power plant accident. Nucl Technol. 1996;114:246–53. https://doi.org/10.13182/NT96-A35253.

3. Kashparov VA, Lundin SM, Khomutinin YV, Kaminsky SP, Levtchuk SE, Protsak VP, Kadygrib AM, Zvarich SI, Yoschenko VI, Tschiersch J. Soil contamination with ^{90}Sr in the near zone of the Chernobyl accident. J Environ Radioact. 2001;56:285–98. https://doi.org/10.1016/S0265-931X(00)00207-1.
4. Kashparov VA, Lundin SM, Zvarich SI, Yoschenko VI, Levtchuk SE, Khomutinin YV, Maloshtan IN, Protsak VP. Territory contamination with the radionuclides representing the fuel component of Chernobyl fallout. Sci Total Environ. 2003;317:105–19. https://doi.org/10.1016/S0048-9697(03)00336-X.
5. Steinhauser G, Brandl A, Johnson TE. Comparison of the Chernobyl and Fukushima nuclear accidents: a review of the environmental impacts. Sci Total Environ. 2014;470–471:800–17. https://doi.org/10.1016/j.scitotenv.2013.10.029.
6. UNSCEAR. Sources and effects of ionizing radiation (annex D). New York: United Nations; 2008.
7. Kashparov V, Levchuk S, Zhurba M, Protsak V, Khomutinin Y, Beresford NA, Chaplow JS. Spatial datasets of radionuclide contamination in the Ukrainian Chernobyl exclusion zone. ESSD. 2018;10:339–53. https://doi.org/10.5194/essd-10-339-2018.
8. Kashparov VA. Hot particles at Chernobyl. Environ Sci Pollut Res. 2003;10(1):21–30. https://doi.org/10.1007/BF02980879.
9. Kuriny VD, Ivanov YA, Kashparov VA, Loshchilov NA, Protsak VP, Yudin EB, Zhyrba MA, Parshakov AE. Particle-associated Chernobyl fall-out in the local and intermediate zones. Ann Nucl Energy. 1993;20(6):415–20.
10. Salbu B, Kashparov V, Lind OC, Garcia-Tenorio R, Johansen MP, Child DP, Roos P, Sancho CM. Challenges associated with the behaviour of radioactive particles in the environment. J Environ Radioact. 2018;186(1):101–15. https://doi.org/10.1016/j.jenvrad.2017.09.001.
11. Kashparov VA, Oughton DH, Zvarich SI, Protsak VP, Levchuk SE. Kinetics of fuel particle weathering and ^{90}Sr mobility in the Chernobyl 30-km exclusion zone. Health Phys. 1999;76:251–9.
12. Kashparov VA, Protsak VP, Ahamdach N, Stammose D, Peres JM, Yoschenko VI, Zvarich SI. Dissolution kinetics of particles of irradiated Chernobyl nuclear fuel: influence of pH and oxidation state on the release of radionuclides in the contaminated soil of Chernobyl. J Nucl Mater. 2000;279:225–33.
13. Kashparov VA, Ahamdach N, Zvarich SI, Yoschenko VI, Maloshtan IN, Dewiere L. Kinetics of dissolution of Chernobyl fuel particles in soil in natural conditions. J Environ Radioact. 2004;72:335–53. https://doi.org/10.1016/j.jenvrad.2003.08.002.
14. Fesenko SV, Alexakhin RM, Balonov MI, Bogdevich IM, Howard BJ, Kashparov VA, Sanzharova NI, Panov AV, Voigt G, Zhuchenka YM. Twenty years' application of agricultural countermeasures following the Chernobyl accident: lessons learned. J Radiol Prot. 2006;26:351–9. https://doi.org/10.1088/0952-4746/26/4/R01.
15. Fesenko SV, Alexakhin RM, Balonov MI, Bogdevitch IM, Howard BJ, Kashparov VA, Sanzharova NI, Panov AV, Voigt G, Zhuchenka YM. An extended critical review of twenty years of countermeasures used in agriculture after the Chernobyl accident. Sci Total Environ. 2007;383(1):1–24. https://doi.org/10.1016/j.scitotenv.2007.05.011.
16. Balonov M, Kashparov V, Nikolaenko E, Berkovsky V, Fesenko S. Harmonization of standards for permissible radionuclide activity concentrations in foodstuffs in the long term after the Chernobyl accident. J Radiol Prot. 2018;38:854–67. https://doi.org/10.1088/1361-6498/aabe34.
17. IAEA. Handbook of parameter values for the prediction of radionuclide transfer in terrestrial and freshwater environments. Technical reports series no. 472. Vienna: IAEA; 2010.
18. Bazyka DA, Tronko MD, Antypkin YG, Serdiuk A, Sushko VO, editors. Thirty years of Chornobyl catastrophe: radiological and health effects: national report of Ukraine, National Academy of Medical Sciences of Ukraine, National Research Centre for Radiation Medicine, Kyiv, Ukraine. 2016. https://drive.google.com/file/d/0B1bUIW1YACgZelRkWmhEMVVIdGc/view.

19. Kashparov V, Levchuk S, Khomutynyn Y, Morozova V, Znurba M. Report of UIAR. Chernobyl: 30 years of radioactive contamination legacy. Kiev: UIAR of NUBiP of Ukraine, Commissioned by Greenpeace Belgium; 2016.
20. Verkhovna Rada of Ukraine. On the legal regime of the territories exposed to radioactive contamination in consequence of the catastrophe at the Chernobyl NPP. Kyiv: Bulletin of Verkhovna Rada; 1991. p. 16.
21. Labunska I, Kashparov V, Levchuk S, Santillo D, Johnston P, Polishchuk S, Lazarev N, Khomutinin Y. Current radiological situation in areas of Ukraine contaminated by the Chernobyl accident: part 1. Human dietary exposure to Caesium-137 and possible mitigation measures. Environ Int. 2018;117:250–9. https://doi.org/10.1016/j.envint.2018.04.053.
22. Maloshtan I, Polishchuk S, Kashparov V, Yoschenko V. Assessment of radiological efficiency of countermeasures on peat-bog soils of Ukrainian Polissya. J Environ Radioact. 2017;175–176:52–9. https://doi.org/10.1016/j.jenvrad.2017.03.026.
23. Lihtarov IA, Kovgan LM, Vasylenko VV. General dosimetry certification and results of whole body counter monitoring in the settlements contaminated after the Chernobyl accident. Data on 2012. Collection 15 (in Ukrainian). Kyiv: Ministry of Health Protection of Ukraine; 2013.
24. Izrael Y, Bogdevich I, editors. The atlas of recent and predictable aspects of consequences of Chernobyl accident on polluted territories of Russia and Belarus (ARPA Russia-Belarus). Moscow-Minsk: Foundation "Infosphere" - NIA-Nature; 2009.
25. UIAR. 2018. www.uiar.org.ua/Eng/nine_milk.htm.
26. Kashparov V, Lazarev N, Polishchuk S. Current problems of agricultural radiology in Ukraine. Agroecol J. 2005;3:31–41.
27. Fesenko S, Jacob P, Ulanovsky A, Chupov A, Bogdevich I, Sanzharova N, Kashparov V, Panov A, Zhuchenka Y. Justification of remediation strategies in the long term after the Chernobyl accident. J Environ Radioact. 2013;119:39–47. https://doi.org/10.1016/j.jenvrad.2010.08.012.
28. Jacob P, Fesenko S, Bogdevitch I, Kashparov V, Sanzharova N, Grebenshikova N, Isamov N, Lazarev N, Panov A, Ulanovsky A, Zhuchenk Y, Zhurba M. Rural areas affected by the Chernobyl accident: radiation exposure and remediation strategies. Sci Total Environ. 2009;408(1):14–25. https://doi.org/10.1016/j.scitotenv.2009.09.006.
29. Ulanovsky A, Jacob P, Fesenko S, Bogdevitch I, Kashparov V, Sanzharova N. ReSCA: decision support tool for remediation planning after the Chernobyl accident. Radiat Environ Biophys. 2011;50:67–83. https://doi.org/10.1007/s00411-010-0344-7.
30. MEU. Twenty-five years after Chornobyl accident: safety for the future. National Report of Ukraine. Ministry of Emergencies of Ukraine. Kyiv: KIM; 2011.
31. Kashparov VA, Protsak VP, Ivanov YA, Nicholson KW. Resuspension of radionuclides and the contamination of village areas around Chernobyl. J Aerosol Sci. 1994;25(5):755–9.
32. Kashparov VA, Protsak VP, Yoschenko VI, Watterson JD. Inhalation of radionuclides during agricultural work in areas contaminated as a result of the Chernobyl reactor accident. J Aerosol Sci. 1994;25(5):761–6.
33. Mamikhin S, Tikhomirov F, Shcheglov A. Dynamics of ^{137}Cs in the forests of the 30-km zone around the Chernobyl nuclear power plant. Sci Total Environ. 1997;193:169–77.
34. Shcheglov A, Tsvetnova O, Klyashtorin A. Biogeochemical cycles of Chernobyl-born radionuclides in the contaminated forest ecosystems. Long-term dynamics of the migration processes. J Geochem Explor. 2014;144:260–6.
35. Thiry Y, et al. Impact of Scots pine (*Pinus sylvestris* L.) plantings on long term ^{137}Cs and ^{90}Sr recycling from a waste burial site in the Chernobyl Red Forest. J Environ Radioact. 2009;100:1062–8.
36. Yoschenko V, Ohkubo T, Kashparov V. Radioactive contaminated forests in Fukushima and Chernobyl. J For Res. 2017;23:1–12. https://doi.org/10.1080/13416979.2017.1356681.
37. Bugai D, Kashparov V, Dewiére L, Khomutinin Y, Levchuk S, Yoschenko V. Characterization of subsurface geometry and radioactivity distribution in the trench containing Chernobyl clean-up wastes. Environ Geol. 2005;47:869–81.

38. Kashparov V, Yoschenko V, Levchuk S, Bugai D, Van Meir N, Simonucci C, Martin-Garin A. Radionuclide migration in the experimental polygon of the red Forest waste site in the Chernobyl zone – part 1: characterization of the waste trench, fuel particle transformation processes in soils, biogenic fluxes and effects on biota. Appl Geochem. 2012;27:1348–58. https://doi.org/10.1016/j.apgeochem.2011.11.004.
39. Evangeliou N, Zibtsev S, Myroniuk V, Zhurba M, Hamburger T, Stohl A, Balkanski Y, Paugam R, Mousseau TA, Møller AP, Kireev SI. Resuspension and atmospheric transport of radionuclides due to wildfires near the Chernobyl nuclear power plant in 2015: an impact assessment. Sci Rep. 2016;6:26062. https://www.nature.com/articles/srep26062.
40. Kashparov V, Zhurba MA, Kireev SI, Zibtsev SV, Myroniuk VV. Evaluation of the expected doses of fire brigades at the Chornobyl exclusion zone in April 2015. Nucl Phys Atom Energy. 2015;16(4):399–407. http://jnpae.kinr.kiev.ua/16.4/Articles_PDF/jnpae-2015-16-0399-Kashparov.pdf.
41. Kashparov V, Myroniuk VV, Zhurba MA, Zibtsev SV, Glukhovskiy AS, Zhukova OM. Radiological consequences of the fire in the Chernobyl exclusion zone in April 2015. Radiat Biol Radioecol. 2017;57(5):512–27.
42. Kashparov V, et al. Forest fires in the territory contaminated as a result of the Chernobyl accident: radioactive aerosol resuspension and exposure of firefighters. J Environ Radioact. 2000;51:281–98.
43. Yoschenko VI, Kashparov VA, Protsak VP, Lundin SM, Levchuk SE, Kadygrib AM, Zvarich SI, Khomutinin YV, Maloshtan IM, Lanshin VP, Kovtun MV, Tschiersch J. Resuspension and redistribution of radionuclides during grassland and forest fires in the Chernobyl exclusion zone: part I. Fire experiments. J Environ Radioact. 2006;86(2):143–63. https://doi.org/10.1016/j.jenvrad.2005.08.003.
44. Yoschenko VI, Kashparov VA, Levchuk SE, Glukhovskiy AS, Khomutinin YV, Protsak VP, Lundin SM, Tschiersch J. Resuspension and redistribution of radionuclides during grassland and forest fires in the Chernobyl exclusion zone: part II. Modeling the transport process. J Environ Radioact. 2006;87(3):260–78. https://doi.org/10.1016/j.jenvrad.2005.12.003.
45. Dewiere L, Bugai D, Kashparov V, Barthès V. Validation of the global model for ^{90}Sr migration from the waste burial in the Chernobyl exclusion zone. Radioprotection. 2005;40(1):S245–51.
46. Levchuk S, Kashparov V, Maloshtan I, Yoschenko V, Van Meir N. Migration of transuranic elements in groundwater from the near-surface radioactive waste site. Appl Geochem. 2012;27(7):1339–47. https://doi.org/10.1016/j.apgeochem.2012.01.002.
47. Kashparov V, Colle C, Zvarich S, Yoschenko V, Levchuk S, Lundin S. Soil-to-plant halogens transfer studies 1. Root uptake of radioiodine by plants. J Environ Radioact. 2005;79(2):187–204. https://doi.org/10.1016/j.jenvrad.2004.06.005.
48. Kashparov V, Colle C, Zvarich S, Yoschenko V, Levchuk S, Lundin S. Soil-to-plant halogens transfer studies 2. Root uptake of radiochlorine by plants. J Environ Radioact. 2005;79(3):233–53. https://doi.org/10.1016/j.jenvrad.2004.07.001.
49. Beresford NA, Barnett CL, Gashchak S, Maksimenko A, Guliaichenko E, Woodb MD, Izquierdo M. Radionuclide transfer to wildlife at a 'reference site' in the Chernobyl exclusion zone and resultant radiation exposures. J Environ Radioact. 2018. (In press). https://doi.org/10.1016/j.jenvrad.2018.02.007.
50. Fuller N, Smith JT, Nagorskaya LL, Gudkov DI, Ford AT. Does Chernobyl-derived radiation impact the developmental stability of *Asellus aquaticus* 30 years on? Sci Total Environ. 2017;576:242–50. https://doi.org/10.1016/j.scitotenv.2016.10.097.
51. Fuller N, Ford AT, Nagorskaya LL, Gudkov DI, Smith JT. Reproduction in the freshwater crustacean *Asellus aquaticus* along a gradient of radionuclide contamination at Chernobyl. Sci Total Environ. 2018;628–629:11–7. https://doi.org/10.1016/j.scitotenv.2018.01.309.
52. Geras'kin S. Ecological effects of exposure to enhanced levels of ionizing radiation. J Environ Radioact. 2016;162–163:347–57. https://doi.org/10.1016/j.jenvrad.2016.06.012.
53. Geras'kin S, Volkova P. Genetic diversity in Scots pine populations along a radiation exposure gradient. Sci Total Environ. 2014;496:317–27. https://doi.org/10.1016/j.scitotenv.2014.07.020.

54. Kashparova E, Levchuk S, Morozova V, Kashparov V. A dose rate causes no fluctuating asymmetry indexes changes in silver birch (*Betula pendula* (L.) Roth.) leaves and Scots pine (*Pinus sylvestris* L.) needles in the Chernobyl exclusion zone. J Environ Radioact. 2018. (In press). https://doi.org/10.1016/j.jenvrad.2018.05.015.
55. Morozova VS, Kashparov VA, Levchuk SY, Umanska AO, Bishchuk YV, Otreshko LM. The functional state of cellular antioxidant defence system of shoots of *Arabidopsis Thaliana* exposed to the chronic ionizing radiation in the Chornobyl exclusion zone. Nucl Phys Atom Energy. 2016;17(3):302–7.. http://jnpae.kinr.kiev.ua/17.3/Articles_PDF/jnpae-2016-17-0302-Morozova.pdf
56. Yoschenko V, Kashparov V, Melnychuk M, Levchuk S, Bondar Y, Lazarev M, Yoschenko M, Farfán E, Jannik G. Chronic irradiation of scots pine trees (*Pinus sylvestris*) in the Chernobyl exclusion zone: dosimetry and radiobiological effects. Health Phys. 2011;101:393–408. https://doi.org/10.1097/HP.0b013e3182118094.
57. Henner P, Hurtevent P, Thiry Y, Levchuk S, Yoschenko V, Kashparov V. Translocation of ^{125}I, ^{75}Se and ^{36}Cl to edible parts of radish, potato and green bean following wet foliar contamination under field conditions. J Environ Radioact. 2013;124:171–84. https://doi.org/10.1016/j.jenvrad.2013.05.012.
58. Hurtevent P, Thiry Y, Levchuk S, Yoschenko V, Henner P, Madoz-Escande C, Leclerc E, Colle C, Kashparov V. Translocation of ^{125}I, ^{75}Se and ^{36}Cl to wheat edible parts following wet foliar contamination under field conditions. J Environ Radioact. 2013;121:43–54. https://doi.org/10.1016/j.jenvrad.2012.04.013.
59. Kashparov V, Colle C, Levchuk S, Yoschenko V, Zvarich S. Radiochlorine concentration ratios for agricultural plants in various soil conditions. J Environ Radioact. 2007;95(1):10–22. https://doi.org/10.1016/j.jenvrad.2007.01.008.
60. Kashparov V, Colle C, Levchuk S, Yoschenko V, Svydynuk N. Transfer of chlorine from the environment to agricultural foodstuffs. J Environ Radioact. 2007;94(1):1–15. https://doi.org/10.1016/j.jenvrad.2006.12.006.
61. Roux C, Le Gal La Salle C, Simonucci C, Van Meir N, Fifield LK, Diez O, Bassot S, Simler R, Bugai D, Kashparov V, Lancelot J. High ^{36}Cl/Cl ratios in Chernobyl groundwater. J Environ Radioact. 2014;138:19–32. https://doi.org/10.1016/j.jenvrad.2014.07.008.
62. Sheppard SC. An index of radioecology, what has been important. J Environ Radioact. 2003;68:1–10.
63. Ivanov V. Likvidatory. Radiologicheskiye posledstviya Chernobylya – Tsentr sodeystviya sotsial'no-ekologicheskim initsiativam atomnoy otrasli. 2010.

Chapter 4
Late Atmospheric Effects of a Nuclear Accident: Comparison Between the Fukushima Daiichi NPP and Chernobyl Accidents

Katsumi Hirose

Abstract Nuclear disasters such as the Chernobyl nuclear power plant (NPP) and Fukushima Daiichi NPP (FDNPP) accidents have contaminated global atmosphere, terrestrial and marine environments by radioactive materials. The environmental impacts of the nuclear accidents continued over more than 10 years. In this chapter, we focus on long-term atmospheric effects of the nuclear accidents by comparing impacts of the FDNPP and Chernobyl for better understanding of their long-term atmospheric effects. For both accidents, the atmospheric concentrations of ^{137}Cs, which is a major radionuclide released from damaged reactors, decreased rapidly with an apparent atmospheric half-life of 1 and 2 weeks at the initial stage, and after that decreased gradually with an apparent atmospheric half-life of about 1 year. The areas affected by the late atmospheric effects correspond to a slow decrease rate of airborne ^{137}Cs, depending on the total release of radioactivity. The late atmospheric effects have been related to radionuclide resuspension and additional emissions from the damaged reactors. However, the current understanding of resupension is more complicated, as it depends on the wind blow of soil particles, human activity in fields as well as on roads and construction sites, forest fires, ecosystem activities of forests, and others. It is noteworthy that a significant fraction of radioactively contaminated areas for both major accidents was forested. These findings suggest that long-term atmospheric radioactivity monitoring is necessary to assess the environmental effects of the nuclear accidents.

Keywords Chernobyl · Fukushima · ^{137}Cs · Late effect · Resuspension · Atmosphere

---

K. Hirose (✉)
Faculty of Science and Technology, Department of Materials and Life Sciences, Sophia University, Tokyo, Japan
e-mail: hirose45037@mail2.accsnet.ne.jp

G. Steinhauser et al. (eds.), *Nuclear Emergencies*,
Current Topics in Environmental Health and Preventive Medicine,
https://doi.org/10.1007/978-981-13-8327-4_4

4.1 Introduction

Two serious nuclear power plant (NPP) accidents, i.e., Chernobyl and Fukushima Daiichi NPP (FDNPP) accidents, have occurred during the past six decades of peaceful use of nuclear energy for electricity production. As a result, huge amounts of radioactivity were released in the environment and spread all over the globe. Radioactivity emitted into the environment has seriously affected human society with potential impacts on human health. The Chernobyl and FDNPP accidents, which have been rated on the International Atomic Energy Agency (IAEA) International Nuclear and Radiological Event Scale (INES) as a "Major Accident" (INES scale 7), were one of the biggest environmental disasters in the recent five decades [1, 2]. In order to implement adequate protective actions for the nuclear disaster and to assess the environmental impact of the Chernobyl/FDNPP radioactivity, a lot of environmental monitoring had been conducted by the national and local governments, research institutes and universities in Russia/Japan and in the world, including the Comprehensive Nuclear-Test-Ban Treaty Organization (CTBTO) network. The major influence of radioactivity released by the NPP accidents appeared for the first 1 month; however, it continued for a long time. In order to predict the impact and fate of possible new NPP accidents to environment, it is important to assess what happened in the past NPP accidents. Therefore, comparison of environmental impacts between the Chernobyl and the FDNPP accidents, which has been recently reviewed [3], is important. In this chapter, we describe the difference in environmental impacts between the Chernobyl and FDNPP accidents, and discuss about similar long-term atmospheric effects, including factors controlling the atmospheric levels of anthropogenic radionuclides.

4.2 Characteristics of Chernobyl and FDNPP Accidents

4.2.1 Overview of the Accidents and Total Releases of Radionuclides

The Chernobyl accident occurred on April 26, 1986, in the course of a technical test in Unit 4 at the Chernobyl NPP, located in Ukraine about 20 km south of the border with Belarus. An initial high atmospheric emission rate of radionuclides on the first day was caused by the explosion of the RBMK-type reactor. There followed a 5-day period of declining releases, which was associated with the hot air and fumes from the burning reactor graphite core materials, after that, the atmospheric release rates of radionuclides increased until tenth day from the initial explosion, and finally the releases stopped sharply. As a result, the radionuclide releases from the damaged reactor occurred mainly over a 10-day period. Major radionuclides released from the Chernobyl accident were due to ^{131}I and ^{137}Cs, taking into account radiological effects and their half-lives. For ^{137}Cs, inventory of the reactor core at the time of accident was estimated to be 260–290 PBq (1 PBq = 10^{15} Bq). The corresponding inventory of ^{131}I was 3200 PBq. The environmental release of ^{137}Cs is estimated to be 85 PBq, corresponding to about 30% of the core inventory. From average deposition densities of ^{137}Cs and the areas of land and ocean regions, the total ^{137}Cs deposition in the northern hemisphere was estimated to be 70 PBq, which is in good agreement with the estimate from the reactor core [1]. For ^{131}I, the release was estimated to be 1760 PBq [1], about 50% of the core inventory. In this case, this value is about five times higher than predictions of the early UNSCEAR (1988) Report [4]. To prevent release of radioactivity in the environment, construction of a sarcophagus covering Unit 4 began in August 1986 and was completed in November 1986.

On March 11, 2011, a 9.0-magnitude earthquake occurred near northeast Honshu, Japan, creating a devastating tsunami. As a result of the earthquake and the subsequent tsunami, the loss of off-site and on-site electrical power (station blackout) and compromised safety systems at the Fukushima Daiichi Nuclear Power Plant (FDNPP) mainly due to flooded diesel electricity generators led to severe core damage to three of the six nuclear reactors on the site [5, 6]. The atmospheric release of radioactivity started in Reactor 1 (BWR MARK I) at night, March 11, 2011, due to melt of nuclear fuel. On March 12, 2011, a hydrogen explosion occurred in the Reactor 1. Large amounts of radioactivity were released in the environment from the FDNPP. On March 14, a hydrogen explosion occurred in the Reactor 3. On March 15, Reactor 2 was seriously damaged. The greatest amounts of radioactivity were released into the atmosphere from March 15 to 16. High emission rates of radioactivity continued until March 23, 2011 [7]. After March 24, emission rates decreased with time.

Major radionuclides released from the FDNPP accident were ^{131}I and ^{137}Cs, as did the Chernobyl accident. For ^{137}Cs, the core inventory of the three reactors at the time of accident was estimated to be 700 PBq. Corresponding core inventory of ^{131}I was 6010 PBq [6]. The atmospheric release of ^{137}Cs was estimated to be 9–36 PBq from reverse and inverse methodologies using monitoring results [2, 7–11]. Aoyama et al. [12] evaluated more accurate total release of ^{137}Cs comparing between model-simulated results including atmosphere and ocean, which was 15–20 PBq, corresponding to about 2% of the core inventory. For ^{131}I, the release is estimated to be 160 PBq, about 3% of the core inventory [6].

For the Chernobyl and FDNPP accidents, major atmospheric emission of radioactivity continued for about 10 days, although the history of radionuclide emission rates differed between the Chernobyl and FDNPP accidents. The total atmospheric release of ^{137}Cs, the most concerned radionuclide from all emissions due to its radiological significance was for the Chernobyl accident by about five times greater than for the FDNPP accident.

4.2.2 Physical and Chemical Properties of Released Radionuclides

The atmospheric behaviors of the radionuclides emitted from the NPP depended on the physical and chemical properties of the radionuclide-bearing particles. For the Chernobyl accident, there were only a few measurements of the aerodynamic diameter of radionuclide-bearing particles released in early days after the accident. A crude analysis of air samples, collected at 400–600 m above the ground in the vicinity of the Chernobyl power plant on April 27, 1987, implied that large radioactive particles, varying the size from several to tens of micrometers, were observed, together with an abundance of smaller particles [13]. The aerosol samples were collected on May 14 and 16, 1986, above the damaged reactor, in which radionuclide-bearing particles showed the superposition of two lognormal distributions: one having an activity median aerodynamic diameter (AMAD) with a range from 0.3 to 1.5 μm, and the other one of more than 10 μm [14]. According to the results of aerosol sampling in remote sites, the AMADs of ^{131}I-, ^{103}Ru-, ^{137}Cs-, and ^{134}Cs-bearing particles were in sub-micrometer range [15–19], whereas the ^{90}Sr and plutonium isotopes were found in larger micrometer particles [20, 21]. The AMADs of Chernobyl-derived radionuclides varied temporally due to the difference in the emission processes at the damaged reactor and/or the fractionation in the transport processes of the Chernobyl radioactivity.

For the FDNPP accident, Doi et al. [22] determined AMADs of particles carrying Fukushima-derived ^{131}I, ^{134}Cs, and ^{137}Cs. The AMAD of ^{131}I-bearing particles was 0.7 μm for both April 4–11 and April 14–21, 2011, events; the AMAD of ^{134}Cs-bearing particles was 1.8 and 1.0 μm in the first and the second period, respectively, while for ^{137}Cs-bearing particles it was 1.5 and 1.0 μm in the first and the second period, respectively. The mass size distribution of the total aerosol was bimodal with peaks in particle diameters at about 0.5 μm and 5–10 μm, which correspond to sulfate and soil particles, respectively. Thus the ^{134}Cs- and ^{137}Cs-bearing particles observed in April differed in diameter from both sulfate and soil particles. The difference in particle size distributions between ^{131}I and radiocesium implies that the process of formation of ^{134}Cs- and ^{137}Cs-bearing particles differed from that of ^{131}I. In another report on the size distribution of Fukushima radiocesium-bearing particles at Tsukuba in the two periods April 28–May 12 and May 12–26, Kaneyasu et al. [23] revealed that ^{134}Cs and ^{137}Cs, having AMAD values around 0.5–0.6 μm, were attached to sub-micrometer sulfate particles. Both findings suggest that the AMAD of the observed radiocesium-bearing particles changed with time. The particle size of radiocesium observed in April may reflect hot particles transported directly from the NPP because the radiocesium concentrations in surface air were more than one order of magnitude greater in April than they were in May. Adachi et al. [24] by using a scanning electron microscope equipped with an energy dispersive X-ray spectrometer revealed that FDNPP-derived radionuclides emitted during the period of March 15–16 were contained in spherical radiocesium-bearing particles (diameter: 2.6 μm), which were water less soluble than sulfate particles. Masson et al. [25] determined size distributions of the FDNPP-derived radionuclide-bearing particles at several places in Europe; the AMAD ranged from 0.25 to 0.71 μm for ^{137}Cs, from 0.19 to 0.69 μm for ^{134}Cs, and from 0.30 to 0.53 μm for ^{131}I, thus in the accumulation mode of the ambient aerosols (0.1–1 μm).

Although the reactor types of the Chernobyl (RBMK) and FDNPP (BWR) differed from each other, most of the ^{137}Cs-bearing aerosols existed as sub-micrometer particles in both cases [15, 23]. At an early stage of the both accidents, significant amounts of radiocesium were emitted into the atmosphere as hot particles, easily removed from the atmosphere by dry and wet deposition processes, unlike the submicrometer particles emitted from Chernobyl [15, 20, 26]. The hot particles derived from the Chernobyl accident were classified into two broad categories: (1) fuel fragments with a mixture of fission products bound to a matrix of uranium oxide, similar to the composition of the fuel in the core, so including plutonium isotopes and other actinides, but sometimes strongly depleted in volatile fission products such as radiocesium, radioiodine, and radioruthenium, and (2) particles consisting of one dominant element (ruthenium and barium) but sometimes also having traces of other elements [27–31]. These monoelemental particles may have derived from embedments of these elements produced in the fuel during the operation and released during the fragmentation of the fuel [28]. For the FDNPP accident, hot particles with spherical shape and amorphous structure contained high amounts of radiocesium, which was embedded into silicate [24, 32]. There is no clear evidence of presence of hot particles consisting of fuel materials. The difference in hot particles between the Chernobyl and FDNPP accidents may be due to the difference in the formation processes of hot particles in the reactor and/or particle formation in release processes [33].

4.2.3 Radioactively Contaminated Areas

In order to effectively conduct radiation protective actions for the FDNPP-derived radionuclide distribution for NPP accidents, it is essential to construct detailed radioactivity contamination (deposition density) maps. After the Chernobyl accident, radioactive contamination of the ground surface was found to some extent in practically every country of the northern hemisphere. The detailed contamination patterns had been established from extensive monitoring of the affected areas. The high contamination area of ^{137}Cs (>37 kBq m^{-2}), which is greater than that of maximum deposition density due to global fallout (~10 kBq m^{-2}) [34], was estimated to be 1.82×10^5 km^2, in which about 75% of the total highly contaminated area are present in the territories of Belarus, the Russian Federation, and Ukraine, where about 25% exists in north and east Europe [1]. The highly ^{137}Cs contaminated areas spread in Belarus (B), the Russian Federation (RF), and Ukraine (U) and were classified as four classes: the class 1 area (>1.48 MBq m^{-2}) was estimated to be 3100 km^2 (RF: 300, B: 2200, U: 600 km^2), the class 2 area (0.555–1.48 MBq m^{-2}) was 7200 km^2 (RF: 2100, B: 4200, U: 900 km^2), the class 3 area (0.185–0.555 MBq m^{-2}) was 19,100 km^2 (RF: 5700, B: 10,200, U: 3200 km^2), and the class 4 area (37–185 kBq m^{-2}) was 116,900 km^2 (RF: 49,800, B: 29,900, U: 37,200 km^2) [1]. The contaminated areas (>185 kBq m^{-2}) in Belarus are 43% agricultural areas, 39% forested, and 2% rivers and lakes.

The highly contaminated area of the FDNPP-derived radionuclides was limited in Japanese territory. The size of the contamination area in Japan with levels >185 kBq m^{-2} after the FDNPP accident, in comparison, is measured by an area of approximately 1700 km^2 [3], which is <6% of corresponding contaminated area for the Chernobyl accident (29,400 km^2). This result is consistent with the findings that about 20–30% of the total atmospheric ^{137}Cs release from the FDNPP was deposited on land [12] and that the total amount of FDNPP-derived ^{137}Cs is about 20% of that of the Chernobyl. In Japan, more than 75% of the contaminated area is forested, <10% rice paddy fields, <10% other agricultural areas, and <5% urban areas.

4.2.4 Atmospheric Effects

Radioactivity measurement in surface air is one of the most important issues in emergency environmental monitoring. After the Chernobyl accident, high levels of radioactivity in surface air and deposition (wet and dry) were observed in early May 1986 at many air monitoring stations in the northern hemisphere. The high activities of Chernobyl radionuclides, typically ^{131}I, ^{137}Cs, ^{134}Cs, and ^{103}Ru, were found in air samples, in which a maximum occurred in early May and after that rapidly decreased with time, although second and third peaks were observed at some sampling stations [35]. An apparent atmospheric half-life of the Chernobyl radionuclides was estimated to be 6 days for observation at remote sites [36]. However, the decrease rate of the Chernobyl radionuclides in air and deposition was slow down. Atmospheric effects of the Chernobyl radioactivity continued for a long time [6].

For the FDNPP accident, various air monitoring campaigns of radionuclides released from the FDNPP had been conducted to elucidate the emission history, which is closely related to sequence of the FDNPP accident, although we had incomplete information on the environmental contamination at the early stage of the accident due to destruction of monitoring systems as a result of the great earthquake and the resulting tsunami [6]. The high level of radionuclides, typically ^{131}I, ^{137}Cs, and ^{134}Cs, was observed in surface air until late March 2011. Although some high peaks of air radionuclides, accompanied with the arrival of the radioactive plume from the FDNPP, occurred during the period of March to April, the level of air radionuclides decreased rapidly, which is corresponding to the decrease of radioactivity emission rates in the FDNPP. An apparent atmospheric half-life of the FDNPP-derived ^{137}Cs during the period of March to June 2011 was calculated to be about 12 days from monthly deposition data [37]. After July 2011, the decrease rates of the surface air concentration and deposition of FDNPP-derived ^{137}Cs declined. In Austria, atmospheric ^{131}I exhibited an apparent half-life between 4 and 6 days and was detectable until May 16, 2011 [38].

4.3 Late Atmospheric Effects of NPP Accidents

4.3.1 Trends of Atmospheric Radionuclides

After the NPP accidents, the terrestrial environment suffered major contaminations, dominantly including forest and agricultural areas. Radioactivity contaminated area is a potential source of radioactive aerosols due to resuspension. On the other hand, sporadic and/or continuous emission of radionuclides from the damaged nuclear reactors may occur in isolation and remediation processes. Therefore, long-term monitoring of anthropogenic radionuclides in air and deposition samples has been required to assess the environmental effects of post-accident.

Since the mid-1950s, radioactivity monitoring sites in Europe, USA, and Japan have been constructed to measure anthropogenic radionuclides, especially ^{137}Cs, ^{90}Sr, and $^{239,240}Pu$ in surface air and deposition to elucidate the effects of nuclear events such as atmospheric nuclear testing, NPP accident, and others. The monitoring results revealed that the ^{137}Cs in surface air and deposition does not return to pre-Chernobyl level at 1 year after the accident. Figure 4.1 shows the temporal variations of annual ^{137}Cs, ^{90}Sr, and $^{239,240}Pu$ deposition in Germany and Greece during the period of 1987–1997 [39–41], in which Tsukuba (Japan) is selected as a reference site [42–44]. In pre-Chernobyl era, the surface air concentrations of anthropogenic radionuclides were controlled by stratospheric–tropospheric fallout due to the atmospheric nuclear testing; in the mid-latitude region, similar level of anthropogenic radionuclides was observed at monitoring stations [45]. In the post-

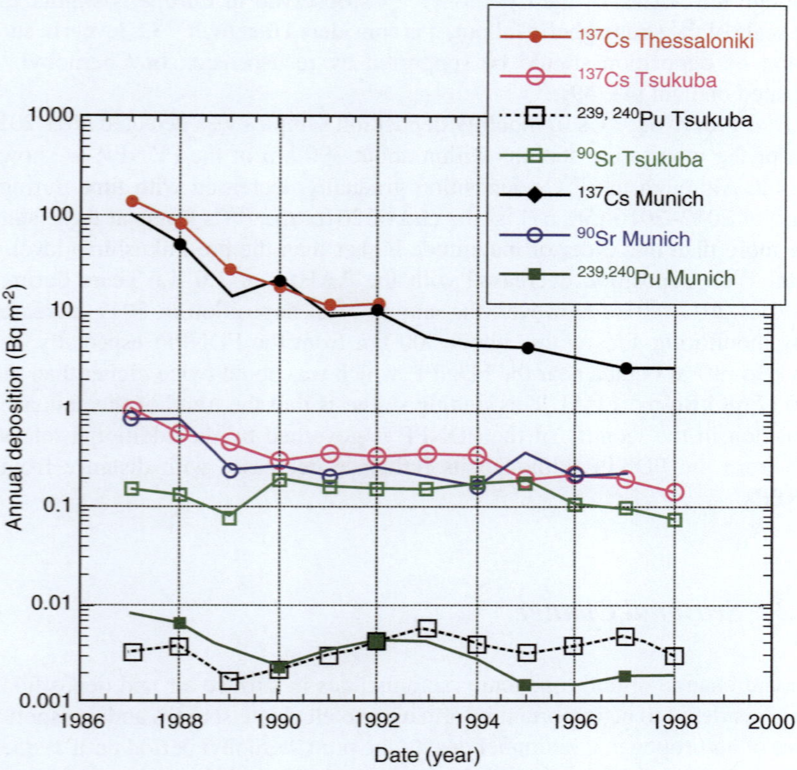

Fig. 4.1 Temporal variations of annual ^{137}Cs deposition observed in Europe and Japan during the period of 1987–1998. The figure was synthesized form papers [39–41]

Chernobyl era, the level of ^{137}Cs deposition in Europe (Germany and Greece) was more than two orders of magnitude larger than that in the Fareast Asia (Japan). The annual depositions of ^{90}Sr and $^{239,240}Pu$ in Munich were affected by the Chernobyl fallout in the late 1980s, although impacts of the Chernobyl-derived ^{90}Sr and $^{239,240}Pu$ are weaker than those of ^{137}Cs. The annual ^{137}Cs deposition in European sites decreased with apparent atmospheric half-life (AAHL) of 1 y; AAHLs at Neuherberg, Germany [46], Mappenberg, Germany [47], and Thessaloniki, Greece [48] were 0.77 y (Aug. 1986–Dec. 1988), 1.07 y (1988–1989), and 1.33 y (1987–1992), respectively. After 1992, the decrease rate of annual ^{137}Cs deposition at Neuherberg was declined. These sites are located at a distance of about 1500 km from Chernobyl. The similar time scale of the AAHL (0.47–2.2 years) was determined for over 20 European sites [49]. The surface ^{137}Cs concentrations in Bratislava were decreasing with the AAHL of 1.9 years during the period of 2003–2010 [50]. On the other hand, irrespective of the long distance of about 8000 km from Chernobyl, small amounts of Chernobyl-derived ^{134}Cs were detected in deposition collected in 1987 at Tsukuba, which may be explained by fallout of the Chernobyl radiocesium partly transported into the lower stratosphere [51–53]. Radiocesium from Chernobyl even significantly elevated the contamination levels in Japanese wheat in 1986 [54]. Although the AAHL of the Chernobyl ^{137}Cs observed in Europe is similar to the time scale of the stratospheric fallout, it is considered that high ^{137}Cs levels in surface air and its deposition should be supported by resuspension of Chernobyl ^{137}Cs deposited on land [41, 49].

FDNPP-derived ^{137}Cs in monthly deposition samples was detected after 2012 at most of the monitoring stations within about 300 km of the FDNPP, as shown in Fig. 4.2. Although the ^{137}Cs deposition gradually decreased with time during the period of 2012–2016 [56, 57], at the end of 2016, the ^{137}Cs levels at these stations were more than one order of magnitude higher than the pre-Fukushima level. The annual ^{137}Cs deposition decreased with the AAHLs of 1.0–1.6 years during the period of 2012–2016. However, the annual ^{137}Cs deposition in 2017 increased at many monitoring sites within about 300 km from the FDNPP; especially, it was 15.3 kBq m^{-2} at Futaba near the FDNPP, which was about twice higher than that in 2016 (7.68 kBq m^{-2}) [55]. This finding suggests that the trend of the annual ^{137}Cs deposition in the vicinity of the FDNPP is governed by the additional release of ^{137}Cs from the FDNPP, although its influence decreases with distance from the FDNPP.

4.3.2 Seasonal Change

Seasonal change of anthropogenic radionuclides in surface air and deposition has been considered to be important information to elucidate sources and transport processes of anthropogenic radionuclides. In the pre-Chernobyl period until 1985, seasonal change of ^{137}Cs in air and deposition, whose peaks occur in May–June, is

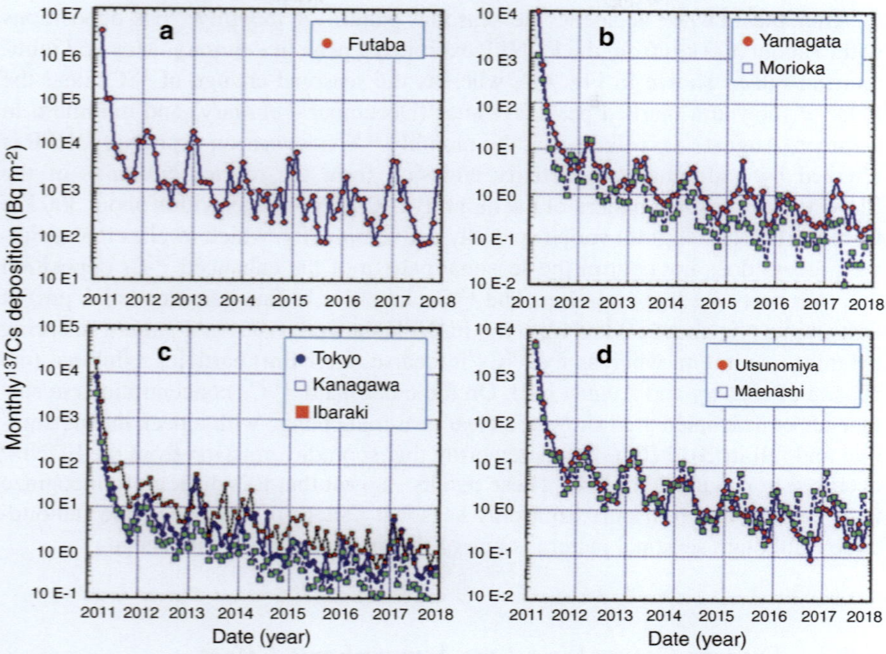

Fig. 4.2 Temporal variations of monthly ^{137}Cs deposition observed in east central Japan. The figure was depicted by using data of NRA [55]. (**a**) Futaba-Okuma near the FDNPP (37.40°N 149.99°E), (**b**) northwest inland sites, Yamagata (38.25°N 140.33°E) and Morioka (39.68°N 141.13°E), (**c**) southwest sites near Pacific coast, Ibaraki-Hitachinaka (36.40°N 140.58°E), Tokyo (35.71°N 139.70°E), and Kanagawa-Chigasaki (35.33°N 139.38°E), (**d**): southwest inland sites, Utsunomiya (36.60°N 139.94°E) and Maehashi (36.40°N 139.10°E)

strongly related to stratospheric fallout due to the atmospheric nuclear testing [58, 59]. According to measurements of ^{137}Cs activity in surface air in Europe, for the post-Chernobyl era, the seasonal pattern of ^{137}Cs exhibits two peaks with enhanced activity in spring (April) and second peak in fall (October). These peaks were explained by advection through the atmosphere boundary layer from Chernobyl. Although seasonal change pattern of surface air ^{137}Cs activity varied between sampling locations (44.5°N–68°N), seasonal change is governed by atmospheric transport from the Chernobyl [59]. On the other hand, the monthly deposition of anthropogenic radionuclides (^{137}Cs, ^{90}Sr, and ^{239,240}Pu) in the 1990s and 2000s exhibited clear seasonal pattern with a spring peak (March–May) in Japan, which is concluded the long-range transport of Asian dust, including global fallout radionuclides without those from Chernobyl, blown up in the east Asian deserts and arid area, based on the knowledge about their level, activity ratios between anthropogenic radionuclides, isotope ratios of plutonium, major element composition of dust including anthropogenic radionuclides, seasonal and inter-annual variations of frequency of dust events and model simulation [43, 44, 60, 61].

After the FDNPP accident, the seasonal pattern of monthly ^{137}Cs depositions within about 300 km from the FDNPP exhibited a peak in February–May and minimum in fall as shown in Fig. 4.2, whereas the seasonal change of ^{137}Cs near the FDNPP showed a marked peak in winter (December–February) and minimum in summer (August). It is likely that the monthly ^{137}Cs deposition near the FDNPP is affected by additional radioactivity emission from the reactor buildings of the FDNPP. The seasonal pattern of the monthly ^{137}Cs depositions within about 300 km from the FDNPP slightly varied spatially and temporally, which implies that a simple process does not control the seasonal pattern of the enhanced ^{137}Cs deposition due to a post-accident emission. The ^{137}Cs activities in fine particles (<1.1 μm) at Fukushima site about 60 km from the FDNPP showed a seasonal pattern in spring (March) maximum, whereas the ^{137}Cs in coarse (>1.1 μm) particles exhibited two peaks in February and August [62]. On the other hand, ^{137}Cs concentrations in surface air of evacuated area showed a clear seasonal change with a peak during summer and fall in 2013 [63]. It is noteworthy that sporadic emission from the FDNPP occurred in August 2013 [64]. These results suggest that it is difficult to recognize their sources and transport from only seasonal pattern of anthropogenic radionuclides, although seasonal pattern is one of the most important knowledge.

4.3.3 Factors Controlling Late Atmospheric Effect

Resuspension is an important process to sustain a level of anthropogenic radioactive aerosols in the surface air. The radionuclides deposited onto ground and/or vegetation are adsorbed onto fine organic or mineral particles. Some meteorological conditions such as aridness and strong wind may blow off fragments of dried soil and vegetation [65]. The resuspension of ^{137}Cs deposited on land surface has been discussed as a significant process supporting the post-accident ^{137}Cs levels in surface air and deposition [66]. Garger et al. [66] summarized the resuspension sources following the Chernobyl accident as (1) dust emission (2) human activity in fields as well as on roads and construction sites, (3) forest fires, and (4) emissions from the power plant (i.e., opening of the Chernobyl sarcophagus). As other resuspension sources, burning of contaminated wastes including biomass and bioaerosols (pollen, spores, bacteria and others) [67, 68] is speculated as the process to support air concentrations and deposition of the FDNPP-derived ^{137}Cs since 2012. Although emission of soil dust is a possible process supporting ^{137}Cs in atmosphere, levels of FDNPP-derived ^{137}Cs deposition observed in Kanto area, central Japan, were hardly supported by dispersion of local soil particles, although its seasonal pattern with high in spring was similar to local dust emission [69]. As a cause of the late atmospheric effects of FDNPP-derived radionuclides, continuous additional atmospheric emission of radiocesium from the FDNPP may be contributed significantly [56].

Model simulation has been developed to assess long-term changes of atmospheric radiocesium after the FDNPP accident, in which includes resuspension from bare soil and forest ecosystem [67, 68], in which a resuspension scheme for

^{137}Cs from bare soil was formulated based on the observation of schoolyard [70] and a forest ecosystem scheme contained term of the green area fraction using normalized difference vegetation index. In this model, resuspension rate from forest was used a constant value of 10^{-7} h. The model simulation reproduced the seasonal pattern of surface atmospheric ^{137}Cs in 2013. However, model simulation of inter-annual trend remains an issue [5].

4.3.4 *Effect of Sporadic Emission*

The radionuclides derived from the Chernobyl accident, as did global fallout, had contaminated wide areas of Eurasia [1] and remain a potential source of radioactive aerosols. Forest fires have frequently happened in Eurasia causing the emission of aerosols. Radiocesium, radioiodine, and chlorine were found in the smoke of biomass fires [71]. The high ^{137}Cs concentrations in surface air sporadically occurred in Lithuania. This was due to the transport of the biomass burning plumes including anthropogenic radionuclides, especially volatile radionuclides such as ^{137}Cs, which led to lower $^{239,240}Pu/^{137}Cs$ ratios than that of global fallout [72]. The radiological risk due to wildfires in Chernobyl contaminated forests was evaluated by model simulation [73]. For the FDNPP accident, there is no clear evidence that biomass burning had a significant impact on atmospheric ^{137}Cs [68].

Sporadic peak events of radioactive emission occurred in August 2013, associated with the debris removal operation in the damaged reactors. High ^{137}Cs concentration in surface air was observed at Namie [67]. Steinhauser et al. [64] estimated the gross release amount of 0.28 TBq to this event using measurements of weekly air filter sampling and monthly deposition and a model simulation.

4.4 Conclusion

The nuclear disasters such as the Chernobyl and FDNPP accidents have caused serious radioactivity contamination of the environment. The environmental impacts of radioactivity released from the nuclear accidents depend on the scale of the total emission from the damaged reactors. After the NPP accidents, atmospheric levels of radionuclides rapidly decreased due to the initial action to cease large emission of radionuclides from the damaged reactors, resulting in apparent atmospheric half-life of 1–2 weeks. A major radionuclide affecting at long time-scale impacts of NPP accidents is ^{137}Cs, although ^{90}Sr and fissile materials such as plutonium are locally important pollutants as well. High atmospheric levels of ^{137}Cs derived from the NPP accident continued over time scale of 1 year, resulting in the apparent atmospheric half-life of about 1 year. However, areas, where late atmospheric effects are observed, depend on spreading of heavily contaminated material, thus ultimately depending on the total radioactivity release. The late atmospheric effect of radioactivity is governed by resuspension of radionuclides initially deposited on land surface, including forests, grass land, agricultural fields, and artificial construction, although additional emission of radioactivity from the damaged reactors occurs continuously and sporadically accompanied with remediation. The current implication of resuspension is complicated because of forests, which is the dominant type of vegetation in the heavily radioactively contaminated areas for both Chernobyl and FDNPP. In order to assess the late atmospheric effects of the NPP accident, it is important to gain better understanding of mechanisms and controlling factors of resuspension, especially host aerosols containing radionuclides.

Acknowledgements The author gratefully acknowledges G. Steinhauser for providing the opportunity to write this chapter and for kind suggestions. The author heartily thanks P.P. Povinec for useful suggestions.

References

1. UNSCEAR. Sources and effects of ionizing radiation volume 2 (Annex J). New York: United Nation; 2000.
2. UNSCEAR. Sources and effects of ionizing radiation (Annex A). New York: United Nation; 2013.
3. Steinhauser G, Brandl A, Johnson TE. Comparison of the Chernobyl and Fukushima nuclear accidents: a review of the environmental impacts. Sci Total Environ. 2014;470-471:800–17.
4. UNSCEAR. Sources and effects of ionizing radiation (Annex D). New York. United Nation; 1988.
5. Hirose K. Fukushima Daiichi nuclear power plant accident: atmospheric and oceanic impacts over the five years. J Environ Radioact. 2016;157:113–30.
6. Povinec PP, Hirose K, Aoyama M. Fukushima accident: radioactivity impact on the environment. New York: Elsevier; 2013.
7. Katata G, Chino M, Kobayashi T, Terada H, Ota M, Nagai H, Kajino M, Draxier R, Hort MC, Malo A, Torii T, Sanada Y. Detailed source term estimation of the atmospheric release for the Fukushima Daiichi nuclear power station accident by coupling simulations of atmospheric dispersion model with improved deposition scheme and oceanic dispersion model. Atmos Chem Phys. 2015;15:1029–70.
8. Chino M, Nakayama H, Nagai H, Terada H, Katata G, Yamazawa H. Preliminary estimation of released amounts of ^{131}I and ^{137}Cs accidentally discharged from the Fukushima Daiichi nuclear power plant into the atmosphere. J Nucl Sci Technol. 2011;48:1129–34.
9. Stohl A, Selbert P, Watowa G, Arnord D, Burkhart JF, Eckhardt S, Tapia C, Vargas A, Yasunari TJ. Xenon-133 and caesium-137 releases into the atmosphere from the Fukushima Dai-ichi nuclear power plant: determination of the source term, atmospheric dispersion, and deposition. Atmos Chem Phys. 2012;12:2313–43.
10. Terada H, Katada G, Chino M, Nagai H. Atmospheric discharge and dispersion of radionuclides during the Fukushima Daiichi nuclear power plant accident. Part II: verification of the source term and analysis of regional-scale atmospheric dispersion. J Environ Radioact. 2012;112:141–54.
11. Winiarek V, Bocquet M, Saunier O, Mathieu A. Estimation of errors in the inverse modeling of accidental release of atmospheric pollutant; application to the reconstruction of the cesium-137 and iodine-131 source terms from the Fukushima Daiichi power plant. J Geophys Res. 2012;117:D05122/1–16.

12. Aoyama M, Kajino M, Tanaka TY, Sekiyama TT, Tsumune D, Tsubono T, Hamajima Y, Inomata Y, Gamo T. ^{134}Cs and ^{137}Cs in the North Pacific Ocean derived from the Mar 2011 TEPCO Fukushima Dai-ichi nuclear power plant accident Japan. Part II: estimation of ^{134}Cs and ^{137}Cs inventories in the North Pacific Ocean. J Oceanogr. 2016;72:67–76.
13. Izrael YA, Vakulovskii SM, Vetrov VA, et al. Chernobyl: radioactive contamination of the environment. Leningrad: Gidrometeoizdat; 1990.
14. Borisov NB, Verbov VV, Kaurov GA, et al. Contents and concentrations of gaseous and aerosol radionuclides above Unit-4 of the Chernobyl NPP. In: Questions of ecology and environmental monitoring, vol 1. Moscow; 1992.
15. Aoyama M, Hirose K, Takatani S. Particle size dependent dry deposition velocity of the Chernobyl radioactivity. In: Precipitation scavenging and atmospheric exchange processes; Fifth International Conference, vol. 3. Washington: Hemisphere Publishing; 1992. p. 1581–93.
16. Bondietti EA, Brantley JN. Characteristics of Chernobyl radioactivity in Tennessee. Nature. 1986;322:313–4.
17. Jost DT, Gaggeler HW, Baltensperger U, Zinder B, Haller P. Chernobyl fallout in size fractionated aerosols. Nature. 1986;324:22–3.
18. Knuth RH, Sanderson CG. Particle size distribution of Chernobyl related aerosols in New York City. In: A compendium of the environmental measurement's research projects related to the Chernobyl nuclear accident. EML-460. New York: US Department of Energy; 1986. p. 291–300.
19. Ooe H, Seki R, Ikeda N. Particle-size distribution of fission products in airborne dust collected at Tsukuba from April to June 1986. J Environ Radioact. 1987;6:219–33.
20. Hirose K. Geochemical studies on the Chernobyl radioactivity in environment. J Radioanal Nucl Chem Artic. 1995;197:315–35.
21. Hirose K, Sugimura Y. Plutonium isotopes in the surface air in Japan: effect of Chernobyl accident. J Radioanal Nucl Chem Artic. 1990;138:127–38.
22. Doi T, Masumoto K, Toyoda A, Tanaka A, Shibata Y, Hirose K. Anthropogenic radionuclides in the atmosphere observed at Tsukuba: characteristics of the Fukushima derived radionuclides. J Environ Radioact. 2013;122:55–62.
23. Kaneyasu N, Ohashi H, Suzuki F, Okuda T, Ikemori F. Sulfate aerosol as a potential transport medium of radiocesium from the Fukushima nuclear accident. Environ Sci Technol. 2012;46:5720–6.
24. Adachi K, Kajino M, Zaizen Y, Igarashi Y. Emission of spherical cesium-bearing particles from an early stage of the Fukushima nuclear accident. Sci Rep. 2013;3:2554. https://doi.org/10.1038/srep02554.
25. Masson O, Ringer W, Malá H, Rulik P, Dlugosz-Lisiecka M, Eleftherisdis K, Meisenberg O, De Vismes-Ott A, Gensdarmes F. Size distributions of airborne radionuclides from the Fukushima nuclear accident at several places on Europe. Environ Sci Technol. 2013;47:10995–1003.
26. Hirose K, Takatani S, Aoyama M. Wet deposition of long-lived radionuclides derived from the Chernobyl accident. J Atmos Chem. 1993;17:61–73.
27. Devell L. Nuclide composition of Chernobyl hot particles. In: Hot particles from the Chernobyl fallout, vol. 16. Ostbayern Theuern: Schriftenreiche des Bergbauunt Industrie-museeums; 1988. p. 23–34.
28. Devell L. Composition and properties of plume and fallout materials from the Chernobyl accident. In: Moberg L, editor. The Chernobyl fallout in Sweden – results from a research programme on environmental radiology. Stockholm: Swedish Radiation Protection Institute; 1991. p. 29–46.
29. Kuriny VD, Ivanov YA, Kashparov VA, et al. Particle-associated fall-out in the local and intermediate zones. Ann Nucl Energy. 1993;20:415–20.
30. Steinhäusler F. Hot particles in the Chernobyl fallout. In: Hot particles from the Chernobyl fallout, vol. 16. Theuern: Schriftenreiche des Bergbau- und Industriemuseums Ostbayerns; 1988. p. 15–21.

31. Steinhäusler F. Summary of the present understanding of the significance of hot particles in the Chernobyl fallout. In: Hot particles from the Chernobyl fallout, vol. 16. Theuern: Schriftenreiche des Bergbau- und Industriemuseums Ostbayerns; 1988. p. 143–4.
32. Satou Y, Sueki K, Sasa K, Adechi K, Igarashi Y. First successful isolation of radioactive particles from soil near the Fukushima Daiichi nuclear power plant. Anthropocene. 2016;14:71–6. https://doi.org/10.1016/j.anene.2016.05.001.
33. Steinhauser G. Anthropogenic radioactive particles in the environment. J Radioanal Nucl Chem. 2018;318:1629–39.
34. Aoyama M, Hirose K, Igarashi Y. Re-construction and updating our understanding on the global weapons tests ^{137}Cs fallout. J Environ Monit. 2006;8:431–8.
35. Povinec P, Chudý M, Sýkora I, Szarka J, Pikna M, Holý K. Aerosol radioactivity monitoring in Bratislava following the Chernobyl accident. J Radioanal Nucl Chem Lett. 1988;126:467–78.
36. Cambray RS, Cawse PA, Garland JA, Gibson JAB, Johnson P, Lewis GNJ, Newton D, Salmon L, Wade BO. Observations on radioactivity from the Chernobyl accident. Nucl Energy. 1987;26:77–101.
37. Hirose K. 2011 Fukushima Daiichi nuclear power plant accident: summary of regional radioactivity deposition monitoring results. J Environ Radioact. 2012;111:13–7.
38. Steinhauser G, Merz S, Kübber-Heiss A, Katzlberger C. Using animal thyroids as ultrasensitive biomonitors for environmental radioiodine. Environ Sci Technol. 2012;46:12890–4.
39. Hirose K, Igarashi Y, Aoyama M. Analysis of 50 years records of atmospheric deposition of long-lived radionuclides in Japan. Appl Radiat Isot. 2008;66:1675–8.
40. Papastefanou C, Ioannidou A, Stoulos S, Manolopoulou M. Atmospheric deposition of cosmogenic ^{7}Be and ^{137}Cs from fallout of the Chernobyl accident. Sci Total Environ. 1995;170:151–6.
41. Rosner G, Winkler R. Long-term (1986 – 1998) of post-Chernobyl ^{90}Sr, ^{137}Cs, ^{238}Pu and $^{239,240}Pu$ concentrations in air, depositions to ground, resuspension factors and resuspension rates in South Germany. Sci Total Environ. 2001;273:11–25.
42. Igarashi Y, Otsuji-Hatori M, Hirose K. Recent deposition of 90Sr and 137Cs observed in Tsukuba. J Environ Radioact. 1996;31:157–70.
43. Igarashi Y, Aoyama M, Hirose K, Miyao T, Nemoto K, Tomita M, Fujikawa T. Resuspension: decadal monitoring time series of the anthropogenic radioactivity deposition in Japan. J Radiat Res. 2003;44:319–28.
44. Igarashi Y, Aoyama M, Hirose K, Povinec PP, Yabuki S. What anthropogenic radionuclides (^{90}Sr and ^{137}Cs) in atmospheric deposition, surface soils and Aeolian dusts suggest for dust transport over Japan. Water Air Soil Poll Focus. 2005;5:51–69.
45. Hirose K, Povinec PP. Sources of plutonium in the atmosphere and stratosphere-troposphere mixing. Sci Rep. 2015;5:15703.
46. Hötzel H, Rosner G, Winkler R. Long-term behaviour of Chernobyl fallout in air and precipitation. J Environ Radioact. 1989;10:157–71.
47. Bachhuber H, Bunzel K. Background levels of atmospheric deposition to ground and temporal variation of ^{129}I, ^{127}I, ^{137}Cs and ^{7}Be in a rural area of Germany. J Environ Radioact. 1992;16:77–89.
48. Ioannidou A, Papastefanou C. Precipitation scavenging of ^{7}Be and ^{137}Cs radionuclides in air. J Environ Radioact. 2006;85:121–36.
49. Garland JA, Pomeroy LR. Resuspension of fall-out materials following the Chernobyl accident. J Aerosol Sci. 1994;25:793–806.
50. Sýkora I, Holý K, Ješkovský M, Müllerová M, Bulko M, Povinec PP. Long-term variations of radionuclides in the Bratislava air. J Environ Radioact. 2017;166:27–35.
51. Aoyama M, Hirose K, Sugimura Y. The temporal variation of stratospheric fallout derived from the Chernobyl accident. J Environ Radioact. 1991;13:103–15.
52. Jaworowski Z, Kownacka L. Tropospheric and stratospheric distributions of radioactive iodine and cesium after the Chernobyl accident. J Environ Radioact. 1988;6:145–50.
53. Kownacka L, Jaworowskii Z. Nuclear weapon and Chernobyl debris in the troposphere and lower stratosphere. Sci Total Environ. 1994;144:201–13.
54. Komamura M, Tsumura A, Kihou N, Kodaira K. ^{90}Sr and ^{137}Cs contamination of wheat produced in Japan. Survey and analysis during the years 1959 through 1995 including the Chernobyl accident. Radioisotopes. 2002;51:345–63. (in Japanese).

55. NRA (Nuclear Regulation Authority). 2018. http://www.nsr.go.jp/activity/monitoring/monitoring2-2.html. Accessed 1 March 2018.
56. Hirose K. Two-years trend of monthly ^{137}Cs deposition observed in Kanto and South Tohoku areas, Japan: effects of the Fukushima Dai-ichi nuclear power plant accident. J Radioanal Nucl Chem. 2015;303:1327–9.
57. Igarashi Y, Kajino M, Zaizen Y, Adachi K, Mikami M. Atmospheric radioactivity over Tsukuba, Japan: a summary of three years of observations after the FDNPP accident. Prog Earth Planet Sci. 2015;2:44. https://doi.org/10.1189/s40645-015-0066-1.
58. Katsuragi Y, Hirose K, Sugimura Y. A study of plutonium fallout in Japan. Pap Meteor Geophys. 1982;33:85–93.
59. Kulan A. Seasonal ^{7}Be and ^{137}Cs activities in surface air before and after the Chernobyl event. J Environ Radioact. 2006;90:140–50.
60. Hirose K, Igarashi Y, Aoyama M, Kim CK, Kim CS, Chang BW. Recent trends of plutonium fallout observed in Japan: plutonium as a proxy for desertification. J Environ Monit. 2003;5:302–7.
61. Lee HN, Igarashi Y, Chiba M, Aoyama M, Hirose K, Tanaka T. Global model simulations of the transport of Asian and Sahara dust: total deposition of dust mass in Japan. Water Air Soil Poll. 2006;169:137–66.
62. Kitayama K, Ohse K, Shima N, Kawatsu K, Tsukada H. Regression model analysis of the decreasing trend of cesium-137 concentration in the atmosphere since the Fukushima accident. J Environ Radioact. 2016;164:151–7.
63. Ochiai S, Hasegawa H, Kakiuchi H, Akata N, Ueda S, Tokonami S, Hisamatsu S. Temporal variation of post-accident atmospheric ^{137}Cs in an evacuated area of Fukushima prefecture: size-dependent behaviors of ^{137}Cs-bearing particles. J Environ Radioact. 2016;165:131–9.
64. Steinhauser G, Niisoe T, Harada KH, Shozugawa K, Schneider S, Synal H-A, Walther C, Christl M, Nanba K, Ishikawa H, Koizumi A. Post-accident sporadic release of airborne radionuclides from the Fukushima Daiichi nuclear power plant site. Environ Sci Technol. 2015;49:14028–35.
65. Hirose K. Radioactive aerosols: tracer of atmospheric processes. In: Colbeck I, Lazaridis M, editors. Aerosol Science. New York: Willy; 2014. p. 441–61.
66. Garger EK, Kuzmenko YI, Sickinger S, Tschiersch J. Prediction of the ^{137}Cs activity concentration in the atmosphere surface layer of the Chernobyl exclusion zone. J Environ Radioact. 2012;110:53–8.
67. Kajino M, Ishizuka M, Igarashi Y, Kita K, Yoshikawa C, Inatsu M. Long-term assessment of airborne radiocesium after the Fukushima nuclear accident: re-suspension from bare soil and forest ecosystems. Atmos Chem Phys. 2016;16:13146–72.
68. Kinase T, Kita K, Igarashi Y, et al. The seasonal variations of atmospheric $^{134,137}Cs$ activity and possible host particles for their resuspension in the contaminated areas of Tsushima and Yamakia, Fukushima, Japan. Prog Earth Planet Sci. 2018;5:12. https://doi.org/10.1186/s40645-018-0171-z.
69. Hirose K. Temporal variation of monthly ^{137}Cs deposition observed in Japan: effects of the Fukushima Daiichi nuclear power plant accident. Appl Radiat Isot. 2013;66:1675–8.
70. Ishizuka M, Mikami M, Tanaka TY, Igarashi Y, Kita K, Yamada Y, Yoshida N, Toyoda S, Satou Y, Kinase T, Ninomiya K, Shinohara A. Use of a size-resolved 1-D resuspension scheme to evaluation resuspended radioactive material associated with mineral dust particles from the ground surface. J Environ Radioact. 2017;166(Pt 3):436–48. https://doi.org/10.1016/j.jenvrad.2015.12.023.
71. Amiro BD, Sheppard SC, Johnson FL, Evenden WG, Harris DR. Burning radionuclide question: what happens to iodine, cesium and chlorine in biomass fires? Sci Total Environ. 1996;187:93–103.
72. Lujanienè G, Aninkevicious V, Lujanas V. Artificial radionuclides in the atmosphere over Lithuania. J Environ Radioact. 2009;100:108–19.
73. Evangeliou N, Balkanski Y, Cozic A, Hao WH, Møller AP. Wildfires in Chernobyl-contaminated forests and risk to the population and the environment: a new nuclear disaster about to happen? Environ Int. 2014;73:346–58.

Chapter 5
Fear of Radiation: A Comparison of Germany and Japan

Frank Rövekamp

Abstract This chapter proposes that the public perception of the danger presented by a national emergency is mediated by cultural factors as well as media bias. The cases of Germany and Japan are examined in the aftermath of the Fukushima nuclear accident. Whereas the people and media remained relatively calm in Japan, where the crisis originated, public concern in faraway Germany was substantial, even triggering an accelerated nuclear phase-out plan there. Possible explanations for these very different reactions, it is argued, may be identified in differing cultural traits, media bias, and vested interests. Some influential Japanese commentators have pointed out the strong tendency of Germans towards emotionality and irrationality, which apparently may be traced back to the romantic era of the late eighteenth century, leading to a rejection of modernity and scientific progress. They also blame German media and press reporting for augmenting these fears and creating an atmosphere of panic. German commentators, on the other hand, criticized Japanese media for being negligent for not reporting on the significant risks posed by the accident. They point to the so-called "nuclear village," the pro-nuclear individuals and institutions including mainstream media. While there is some debate regarding the degree to which the "village members" actually do connect and deliberately plan, there is a general consensus that the nuclear industry commands strong lobbying powers. The chapter concludes that while both viewpoints may be exaggerated, they do contain a kernel of truth. On an individual level, "fear of radiation" appears to be a very human and explicable impulse independent of a cultural explanation. Collective fear bordering on hysteria of people at a safe distance from the site of an emergency, however, is another matter. Here cultural factors and the media seem to play a vital role.

Keywords Fukushima Daiichi nuclear accident · Media reporting · Romantic mindset · "Nuclear Village" · Japan · Fear

---

F. Rövekamp (✉)
East Asia Institute, Ludwigshafen University of Applied Sciences, Ludwigshafen, Germany
e-mail: roevekamp@oai.de

G. Steinhauser et al. (eds.), *Nuclear Emergencies*,
Current Topics in Environmental Health and Preventive Medicine,
https://doi.org/10.1007/978-981-13-8327-4_5

5.1 Introduction

The Fukushima nuclear accident of March 11, 2011, drew worldwide attention even months after the event and had far-reaching consequences for the international nuclear industry. The reactions, however, were different from country to country. As is to be expected, they were very marked in the country of origin, Japan. But still profounder were the reactions and consequences in a country virtually at the other end of the world, in Germany. How can such a difference be explained? Are cultural factors at work? Is "fear of radiation" an issue of mentality? Or are other factors such as media biases or industry lobbying important?

This chapter is not an attempt to deal with these questions in a comprehensive and fundamental way. But it will try to shed some cross-cultural light on these issues by introducing a Japanese view of German behavior and a German view of the driving forces behind Japanese attitudes. To provide an analytic base, the chapter will first review important public reactions to and the consequences of the nuclear accident in both countries. It will conclude with some reflections on the differences between German and Japanese perspectives.

5.2 Public Reactions During the Fukushima Nuclear Accident

5.2.1 Japan

In Japan itself, the reactions to the nuclear disaster cannot be separated from the causes of the catastrophe, the Great Tohoku Earthquake and the ensuing tsunamis, which devastated large parts of the coastal area in East Japan, cost close to 20,000 lives and displaced several 100,000 people [1, pp. 781–782]. While media attention was gradually drawn to the unfolding nuclear accident, it nevertheless did not divert from extensively reporting about the larger scale earthquake disaster and its enormous human consequences.

Even with the worsening of the situation at the nuclear plant, obvious problems with emergency responses, manifested by the chaotic step-by-step evacuation process[1] and a growing sense of crisis, media reporting remained rather matter-of-fact and did not engage in wide-ranging speculation about core meltdown and worst-case scenarios [3, pp. 257–260]. This had the effect that the situation in areas not directly affected by the accident, like Tokyo and its surroundings, remained rather calm and without panic reactions on a large scale. This did not change even with the spectacular hydrogen explosions at the Fukushima Daiichi reactor buildings on

[1] The Investigation Commission of the National Diet of Japan gives a very detailed account of the evacuation process and its flaws [2, pp. 6–38].

March 12, 14, and 15 or with the evacuation of foreign embassies and other institutions from the Tokyo area to other parts of the country.

There were, of course, differences of tone and degree of detail in the nuclear accident reporting by major television stations and newspapers. The influential national public television channel NHK had the calmest approach. News about the nuclear accident was reported along with other conventional news and speculations about core meltdown and worst-case scenarios were mostly avoided. Major national newspapers differed somewhat in their reporting in line with their principal political orientation. The traditionally conservative and pro-atomic energy Yomiuri Shimbun criticized the government for its chaotic emergency response, but avoided reporting about the accident in a manner that could lead to questions about the use of atomic energy as such. The more liberal Asahi Shimbun and Mainichi Shimbun were somewhat more outspoken in their crisis and risk reporting, but also they avoided engaging in speculative scenarios and assessments that might arouse widespread panic [3, pp. 249–260].

5.2.2 *Germany*

Media reporting and public reactions were entirely different on the other side of the world, in Germany. As soon as it became clear that the nuclear accident was of a very serious nature, the topic commanded virtually exclusive attention. The grave effects of the earthquake and the tsunamis played all but no role in the news anymore [4, pp. 139–144].

Speculation about the extent of the accident and worst-case scenarios was widespread. Experts, some of them well qualified and some of them perhaps less so, mostly from environmental and anti-nuclear groups were present on virtually all news programs. Frequent comparisons with the nuclear accident of 1986 at Chernobyl were made. Many commentators claimed that "Fukushima" could only be worse given the much higher radioactive inventory at the plant. Although unconfirmed at that time it was generally assumed as given that a core meltdown in the nuclear reactors was well under way. The issue discussed then was when the capital area Tokyo would be hit with a radioactive plume and what an evacuation of about 50 million people would look like, if possible at all.[2] Beyond that, the extent Germany would be affected by the accident and how to defend against this "threat" was of great concern. Geiger counters were sold out in a short period of time and iodine tablets were in high demand [9].

A prominent, if somewhat extreme example of German media reporting was the front page headline "Tokyo in mortal fear" ("Tokio in Todesangst") in DIE WELT, a leading newspaper, on March 16 [8]. The headline was above a large picture depicting people wearing masks. This was apparently meant to suggest that the mask wearing was the result of alleged radioactive contamination and widespread

[2] Examples from prominent German media are Focus [5], SZ [6], and WELT [7, 8].

fear in the capital. In fact however Japanese people wear masks very frequently, already with light symptoms of a cold or even as a protective measure in cold and hay fever seasons. Thus the picture was of a very ordinary daily-life scene in Japan and it had no connection with the nuclear accident and its alleged effects.

Shortly after the accident, Germany—which has never experienced a serious nuclear incident within its borders and which is blessed by a relatively calm natural environment with no regular earthquakes or weather extremes—shut down some of its oldest nuclear power plants. Albeit they had still been running without operating problems and they posed no apparent imminent safety risks, the government decided to completely abandon nuclear energy by the year 2022 [10], by an accelerated phase-out plan.

In earthquake-, tsunami-, and typhoon-prone Japan, on the other hand, the discussions on nuclear energy took a complicated turn. Shortly after the accident, the government, under the control of the Democratic Party, decided to phase out nuclear energy between the years 2030 and 2040 [11, pp. 405–444]. This was changed, however, by the government of the Liberal Democratic Party, which came to power again in 2012. The official policy today is to rely continuously on nuclear power to provide a significant base load of electricity [12].

Why did Germans, who had not been directly affected by the accident, react so differently from the Japanese, who were in the middle of the disaster?

5.3 A Japanese View of the German Reaction

The German reaction to the Fukushima nuclear accident was critically analyzed by a Japanese journalist, Norihide Miyoshi, in his prize-winning book "The German Risk" [13]. Miyoshi has a profound knowledge of German affairs. He had lived in Germany for more than 10 years as the foreign correspondent of the Yomiuri Shimbun, the biggest newspaper in Japan and one of the biggest in the world. Miyoshi's book deals with general political culture in Germany, but it starts with media reporting on Fukushima and the ensuing German energy transition.

Miyoshi criticizes the very speculative and emotional tone in general media reporting. Beyond that he points out that this was accompanied by sometimes implicit and sometimes explicit accusations that the Japanese media were colluding with the authorities and not telling the truth about the extent of the disaster and the accompanying risks.[3] German media allegedly "knew better."

At the end of the day, the radioactive fallout from the Fukushima accident amounted to about 10–20% of the fallout from the Chernobyl reactor explosion [15]. Furthermore there were no direct, radiation-related fatalities. Miyoshi contrasts these facts with a Facebook post by Claudia Roth, a prominent German Green Party politician. Roth mixed erroneously the tsunami disaster and the nuclear accident and suggested the latter claimed about 16,000 victims [13, pp. 35–36]. In

[3] See the article in the *Frankfurter Rundschau* [13] as an example for such reporting.

Miyoshi's view, the German reaction to the accident was internationally unmatched in its emotional and smart-alecky allegations made from self-claimed high moral ground.

What are the roots of this behavior? According to Miyoshi, many Japanese assume that the Chernobyl disaster of 1986, which indeed affected Germany and its agriculture through the high level of radioactive fallout, is the main factor behind German reactions. He points out that the fiercely anti-nuclear modern German environmental movement took off in the late 1960s and became an important political force in the 1970s, long before the Chernobyl accident. His main proposition, however, is that German mentality as such is characterized by a deep skepticism towards science, technology, and modernity with roots in the Romantic Era of the nineteenth century [13, pp. 231–242]. According to Miyoshi, Romantic thought has not only permeated arts and music but also strongly influenced political thought up to the present day.[4]

The Romantic Era began at the end of the eighteenth century and was characterized by a mystical mindset and ascribed mystical power to nature. Emotionality was stressed over rationality. Science and technology were strongly resented, and some even rejected "modernity" altogether. As a genre of arts and literature, it declined as soon as the first part of the nineteenth century, but Romantic thought has deeply influenced social and political movements well beyond the Romantic Era even up to the present. Examples include the Youth Movement of the early twentieth century and also parts of the National Socialist movement with its countryside ideology. After World War II, Romantic thought continued to influence phenomena such as the student revolt of the 1960s and the environmental movement, which took off in every respect in the 1970s and which formed the basis of the Green Party. Miyoshi is keen to point out that only in Germany the environmental movement has gained so much strength as to crystallize into a major political party which could even participate in the national government.

Miyoshi's critique thus is a very fundamental one. He sees German reporting of and reactions to the Fukushima nuclear accident as a symptom of a very fundamental cultural trait which is characterized by a deep-rooted skepticism against science and technology or even "modernity" as such.

5.4 A German View of the Japanese Nuclear Village

Miyoshi's book received the prestigious Yamamoto Shichihei prize in Japan for excellent non-fictional writing. Also, his views are influential as a senior staff writer for the Yomiuri Shimbun.

The Yomiuri Shimbun can be considered part of the so-called Nuclear Village in Japan, sobriquet for an informal if not silent coalition of people and institutions

[4] In his chapter about Romanticism, Miyoshi draws heavily from an influential book on this matter by the German philosopher and author Rüdiger Safranski [15].

which strongly promote nuclear energy and profit from it. These are the major utilities, the biggest of them TEPCO, the operator of the Fukushima nuclear plant, local communities which profit from high subsidies for hosting nuclear power plants, pro-nuclear politicians, scientists, and large parts of the major media [17].

The Yomiuri Shimbun plays a very special role in this connection. It is controlled by the Shōriki family and it was Matsutarō Shōriki, who in the 1950s, together with Yasuhiro Nakasone, later prime minister, was the main driving force behind the introduction of nuclear energy into Japan. Shōriki also became chairman of the newly established Japanese Atomic Energy Commission [18, pp. 235–239]. The formidable task of convincing the Japanese public, which still had a vivid memory of the atomic bombings of Hiroshima and Nagasaki, that nuclear energy has nothing to do with nuclear weapons, that it indeed presents the "good power" of atoms, was undertaken by the major newspapers with the Yomiuri Shimbun leading the way [19, pp. 65–110].

The Yomiuri Shimbun has profited greatly from this stance up to the present day. All the ten major utilities have extraordinarily high PR budgets, and regular advertisements in the major newspapers are an important source of income for them [20]. The nuclear accident thus posed a serious threat to the business model of the Yomirui Shimbun and similar media giants. Did this also influence their reporting during the accident and did they underrate the risks? Is there, in other words, substance to the claim of many German media that their Japanese counterparts did not report objectively and thus put the Japanese population at risk?

One indicator for this may be the official American reaction to the accident and how the Japanese side dealt with it. US officials announced a safety zone of 50 miles around the nuclear plant, which all American citizens should vacate [21, p. 80], an announcement largely ignored by Japanese media. Miyoshi, for his part, plays down the seriousness of the accident by declaring that the standard for judging it should be radioactive fallout, which only reached the level of 10–20% of that from Chernobyl. But this final outcome does ignore the fact that things had been on the verge of becoming much worse. One of the reactors could well have burst and in the worst case—outlined by Shunsuke Kondō, the former chairman of the Japan Atomic Energy Commission in March 2011 [22]—eventually an evacuation zone of 250 km around the plant might have become necessary.[5] This would have included the Tokyo area and would have affected around 50 million people. Were Japanese media negligent by not reporting or at least speculating about such a risk or were they acting responsibly by confining their reporting strictly to the confirmed facts and thereby avoiding overwhelming, uncontrollable public panic? Opinions on this remain divided up to the present day.

[5] How easily matters could have developed in this direction was also testified by Masao Yoshida, the Fukushima nuclear plant manager [22, pp. 180–185] and by Haruki Madarame, head of the Nuclear Safety Commission [23, pp. 25–27] at the time of the accident.

5.5 Conclusion

The Japanese "Nuclear Village," its influence and its networks, is an intensively discussed topic in Japan and also in Japan-expert circles outside of the country. There are different opinions about the degree to which the constituents of the "Village" really connect and collude consciously, but there are few doubts that the nuclear industry commands strong lobbying powers. And there is no doubt about the role of the influential mainstream media, with the Yomiuri Shimbun in the lead, promoting nuclear energy as a cheap, reliable and—if managed with sufficient care—safe source of energy.

Miyoshi may thus be considered as a "man of the system," someone who certainly feels close to the "Nuclear Village" and its agenda. But does this discredit his analysis of German media reporting about the Fukushima nuclear accident, public reactions, and its deeper roots? This may not be the case. His analysis of media reporting is thorough and backed up by many sources, and even if counterexamples of more calm and balanced reporting can be detected, one can hardly escape the impression that emotion and a know-it-all attitude were among the driving forces in German media coverage during those days.[6] Miyoshi contrasts this also with the more balanced reporting of important British media like the BBC and the Financial Times [13, pp. 37–41]. As for the long-term consequences of abandoning nuclear energy and enforcing an "energy transition," no other country in the world acted with a speed and determination similar to that of Germany, for better or worse. Whether all this is a symptom of deeply ingrained German cultural traits, characterized by anti-scientific sentiments, pessimism, and a rejection of modernity, with roots in the Romantic Era or even beyond, is certainly debatable. One should note, however, that influential authors like the German philosopher Rüdiger Safranski [16], the late American historian Gordon Craig [26, pp. 216–239], and even the great writer Thomas Mann [27] share or shared Miyoshi's view of the influence of the dark side of Romantic thought on the German soul.

On an individual level, "fear of radiation" appears to be a very human impulse with no relation to age, gender, and race. One needs only to consider the reactions of those directly affected during and shortly after the Fukushima nuclear accident or follow the discussions between TEPCO and local fishermen about releasing contaminated water into the ocean today. Here the Japanese people reacted no differently than would be expected from German or any other people in the same situation. The nature of radioactivity with probable and not easy to understand but nevertheless potentially very serious effects on human health let this appear quite plausible. Human beings, no matter their backgrounds, fear uncertainty and the unknown, if directly confronted with it.

The differing collective reactions of people however, who are at a safe distance from an event, as compared with those on the scene, are another matter. Here

[6] This conclusion can also be drawn from an analysis by German media experts, who linguistically compare German reporting with that in the USA and Great Britain [24].

disparate media and mentalities seem to play an important role. In the Fukushima nuclear accident case, these factors apparently manifested themselves in a sort of complacency or conformity in Japan as contrasted with an emotionally driven overreaction in faraway Germany.

References

1. Kazama M, Noda T. Damage statistics (summary of the 2011 off the Pacific Coast of Tohoku earthquake damage). Soils Found. 2012;52(5):780–92. https://www.sciencedirect.com/science/article/pii/S0038080612000947. Accessed 27 Sept 2018.
2. NAIIC (The Fukushima Nuclear Accident Independent Investigation Commission). Main report. Tokyo: The National Diet of Japan; 2012. http://warp.da.ndl.go.jp/info:ndljp/pid/3856371/naiic.go.jp/en/report/. Accessed 2 Sept 2018.
3. Weiß T. Die japanischen Medien und die Atomkatastrophe von Fukushima. In: Chiavacci D, Wieczorek I, editors. Japan 2014 – Politik, Wirtschaft und Gesellschaft. München: Iudicium; 2014. p. 245–69.
4. Kepplinger HM, Lemke R. Framing Fukushima – Zur Darstellung der Katastrophe in Deutschland im Vergleich zu Großbritannien, Frankreich und der Schweiz. In: Wolling J, Arlt D, editors. Fukushima und die Folgen – Medienberichterstattung, Öffentliche Meinung, Politische Konsequenzen. Ilmenau: Universitätsverlag; 2014. p. 125–52.
5. Focus. Warum eine Evakuierung von Tokio unmöglich ist. 2011. https://www.focus.de/panorama/welt/tsunami-in-japan/hintergrund-warum-eine-evakuierung-von-tokio-unmoeglich-ist_aid_608862.html. Accessed 29 Sept 2018.
6. SZ (Süddeutsche Zeitung). Wehe wenn der Wind dreht. 2011. https://www.sueddeutsche.de/wissen/japan-evakuierung-aus-tokio-wehe-wenn-der-wind-dreht-1.1072862. Accessed 29 Sept 2018.
7. WELT. 35 Millionen Menschen in Gefahr, "Evakuierung unmöglich". 2011. https://www.welt.de/vermischtes/weltgeschehen/article12831993/35-Millionen-Menschen-in-Gefahr-Evakuierung-unmoeglich.html. Accessed 27 Sept 2018.
8. WELT. Tokio in Todesangst. 2011. https://www.welt.de/print/die_welt/politik/article12841234/Tokio-in-Todesangst.html. Accessed 27 Sept 2018.
9. FAZ (Frankfurter Allgemeine Zeitung). Sake unbedenklich, Geigerzähler ausverkauft. 2011. http://www.faz.net/aktuell/rhein-main/region-und-hessen/folgen-der-nuklearkatastrophe-in-japan-sake-unbedenklich-geigerzaehler-ausverkauft-1612206.html. Accessed 29 Sept 2018.
10. Bundesregierung. Energiewende – Fragen und Antworten. 2011. https://www.bundesregierung.de/Webs/Breg/DE/Themen/Energiewende/Fragen-Antworten/8_Kernkraft/_node.html. Accessed 29 Sept 2018.
11. Kushida KE. The DPJ's political response to the Fukushima nuclear disaster. In: Kushida KE, Lipscy PY, editors. Japan under the LDP – the politics of transition and governance. Stanford: Shorenstein APARC; 2013. p. 405–44.
12. METI (Ministry of Economy, Trade and Industry). Cabinet decision on the new strategic energy plan. 2018. http://www.meti.go.jp/english/press/2018/0703_002.html. Accessed 2 Sept 2018.
13. Miyoshi N. Doitsu risuku – yume wo miru seiji ga hikiokosu konran [The German risk – on the chaos of politics of dreaming]. Tokyo: Kōbunsha; 2015.
14. FR (Frankfurter Rundschau). Japanische Medien und Fukushima – Skandal auf Seite 17. 2011. http://www.fr.de/kultur/netz-tv-kritik-medien/medien/japanische-medien-und-fukushima-skandal-auf-seite-17-a-924259. Accessed 29 Sept 2018.
15. Steinhauser G, Brandl A, Johnson TE. Comparison of the Chernobyl and Fukushima nuclear accidents: a review of the environmental impacts. Sci Total Environ. 2014;470-471:800–17.

16. Safranski R. Romantik – Eine deutsche Affäre. München: Carl Hanser; 2007.
17. Kingston J. Japan's nuclear village. Asia Pac J. 2012;10(37):1. https://apjjf.org/-Jeff-Kingston/3822/article.pdf. Accessed 27 Sept 2018.
18. Samuels RJ. The business of the Japanese state – energy markets in comparative and historical perspective. Ithaca: Cornell University Press; 1987.
19. AS (Asahi Shimbun Shuzaihan). Kore demo Nihonjin wa Genpatsu wo eranda [Dennoch haben die Japaner die Atomkraft gewählt]. Tokyo: Asahi Shimbun Shuppan; 2014.
20. Nakano Y. Genpatsu sangyō no kane to hito [Geld und Menschen der Atomindustrie]. 2015. https://ci.nii.ac.jp/els/contentscinii_110009464261.pdf?id=ART0009935137. Accessed 27 Sept 2018.
21. Lochbaum D, Lyman E, Stranahan SQ. Fukushima – the story of a nuclear disaster. New York: The New Press; 2014.
22. Kondō S. Fukushima Daiichi Genshiryokuhatsudensho no Fusokujitai Shinario no Sobyō [A roughly sketched contingency plan for the Fukushima Daiichi Nuclear Power Plant]. 2011. www.asahi-net.or.jp/~pn8r-fjsk/saiakusinario.pdf. Accessed 31 Oct 2017.
23. Kadota R. Yoshida chōsho wo yomikaku [Interpreting the Yoshida protocolls]. Tokyo: PHP; 2014.
24. Okamoto K. Shōgen: Madarame Haruki [Testimony by Haruki Madarame]. Tokyo: Shinchōsha; 2012.
25. Zeller F, Dorothee A, Wolling J. Emotionalisierte Berichterstattung? Wie die Presse in Großbritannien, den USA und Deutschland über die Folgen des Erdbebens und die Tsunamis in Japan berichtete. In: Wolling J, Arlt D, editors. Fukushima und die Folgen – Medienberichterstattung, Öffentliche Meinung, Politische Konsequenzen. Ilmenau: Universitätsverlag; 2014. p. 183–219.
26. Craig GA. Über die Deutschen. München: dtv; 1985.
27. Mann T. Rede über Deutschland und die Deutschen. Berlin: Suhrkamp; 1947.

Chapter 6
The Psychosocial Consequences of the Fukushima Disaster: What Are We Suffering From?

Masaharu Maeda, Yuliya Lyamzina, and Akiko Ito

Abstract The Fukushima Daiichi Nuclear Power Plant accident caused enormous damage in terms of not only the mental status of affected people, but also the cohesiveness of entire communities in Fukushima Prefecture. Regarding individual mental health, many psychiatric issues became apparent after the accident, including, but not limited to, posttraumatic stress disorder, depression, and alcohol or another type of substance abuse. Widespread rumors and damaged reputations caused anxiety among residents and evacuees, eliciting various disparities such as risk perception factors related to compensation or the effects of radiation exposure. As a result, a decrease in community resilience was observed. Additionally, evacuees were frequently exposed to public stigmas resulting from the negative stories regarding compensation issues or the possible genetic effects of radiation exposure. To address these multidimensional mental health problems, several new and unique care facilities were established after the disaster with the aim of providing active interventions for and improving the current well-being of affected people, including evacuees. While a certain level of effectiveness in the provision of outreach services has been seen, issues such as burnout and exhaustion among health care staff working for different care resources have also been observed. In contrast to natural disasters, nuclear disasters tend to have long-term psychosocial consequences on affected people. Therefore, support care resources that could play important roles, especially in the post-disaster phase in affected areas, should be supported by national and local governments on a long-term basis.

Keywords Nuclear disaster · Posttraumatic accident · Depression · Stigma · Risk communication

---

M. Maeda (✉) · Y. Lyamzina · A. Ito
Department of Disaster Psychiatry, School of Medicine, Fukushima Medical University, Fukushima, Fukushima Prefecture, Japan

Radiation Medical Science Center for the Fukushima Health Management Survey, Fukushima Medical University, Fukushima, Fukushima Prefecture, Japan
e-mail: masagen@fmu.ac.jp

G. Steinhauser et al. (eds.), *Nuclear Emergencies*,
Current Topics in Environmental Health and Preventive Medicine,
https://doi.org/10.1007/978-981-13-8327-4_6

6.1 Introduction

On March 11, 2011, a massive earthquake and tsunami struck the Tōhoku region of Japan, resulting in serious damage, especially across Iwate, Miyagi, and Fukushima Prefectures. Furthermore, the tsunami crippled the cooling system at the Fukushima Daiichi Nuclear Power Plant (FDNPP), leading to serial explosions in all four reactor buildings. This subsequent nuclear accident caused not only direct physical loss and damage, but also long-term psychosocial effects on residents across the entire Fukushima region.

Within a month after the accident, about 140,000 people were voluntarily or involuntarily evacuated to different locations inside and outside of Fukushima Prefecture. The evacuees aimlessly traveled from place to place in an attempt to avoid exposure to radiation. It has currently been estimated that the evacuees changed their location approximately four times on average during the first year after the disaster [1]. In the acute phase of the accident, most of the evacuees believed that their relocation was only tentative and that they would be able to return to their homes soon, possibly within several weeks or months. However, the information conveyed by the Japanese government during the acute phase of the accident was often vague and unclear, and this was amplified by differences in opinions among numerous "experts" toward the possible adverse health effects of radiation exposure. Inaccurate and insufficient information about the radiation risk made an already bad situation even worse. Moreover, various additional factors that were originally unrelated to the health effects of radiation exposure, such as political opinions toward nuclear policy in Japan, contributed to the initial confusion and uncertainty among evacuees and those who had decided to stay in the Fukushima area.

Looking back at the situation, the people affected by the Fukushima disaster seemed to be suffering from the effects of baseless rumors and accusations and unfounded suspicions fostered by the so-called experts rather than from the direct effects of the tsunami and subsequent nuclear accident. Information on the internet had an immensely negative effect on the evacuees' judgements and decisions regarding whether to return home and helped shape negative public attitudes and stereotypes about those affected by the disaster as well as the general image of the Fukushima area and its products both inside and outside Japan. A recent report published by a well-known think tank in Japan [2] revealed that over 40% of the people living in Tokyo were still worried about the possible adverse genetic health effects of radiation exposure from the Fukushima accident; many of the respondents reported obtaining their information and knowledge from the internet only, rather than from Japanese governmental sources, scientific publications, or other official sites such as academic societies or international organizations, such as the International Atomic Energy Agency (IAEA), the United Nations Scientific Committee on the Effects of Atomic Radiation (UNSCEAR), and the International Commission on Radiological Protection (ICRP), which are supposed to play major roles in safety and protection against ionizing radiation after rare events such as major nuclear accidents [3].

In this chapter, we first describe the various types of psychological consequences that emerged after the Fukushima disaster, such as posttraumatic stress disorder (PTSD), grief, and loss reactions. Second, we examine the psychosocial issues evoked among evacuees and affected communities, such as the negative influence of mass media and the resultant stigmas and self-stigmas. Third, to identify the multi-dimensional stressors affecting numerous survivors of the Fukushima disaster, a concrete example is provided. Finally, we briefly discuss some of the current important challenges in Fukushima Prefecture that remain to be met. It is our sincere hope that the lessons that can be learned from the Fukushima disaster will be helpful after any potential nuclear crisis in the future.

6.2 Current Psychiatric Issues in Fukushima Prefecture

6.2.1 Posttraumatic Responses and Depressive Symptoms

In 2011, people living in Fukushima Prefecture were exposed to an unfortunate series of complex traumatic events: a massive earthquake followed by a historic tsunami and subsequent nuclear disaster. These events caused substantial trauma among those living in the affected area, such as evacuees, leading to PTSD in many, the symptoms of which include but are not limited to the avoidance of places, people, and certain activities, emotional numbness or re-experiencing the traumatic event, for example, through flashbacks and nightmares followed by hyperarousal and altered arousal responses.

In addition to numerous posttraumatic symptoms, persistent fear and worry related to radiation exposure and food and water contamination negatively affected general public opinion, with the strongest fears and anxieties still prevailing both inside and outside of Fukushima Prefecture, especially among women and mothers [4]. Despite the tremendous human and financial costs, as well as the dedicated resources to the recovery process that were supplied by the Japanese government [5] and the gradual lifting of evacuation orders and reopening of a number of affected areas, many evacuees were still hesitant to return to their hometowns because of continuous groundless rumors, baseless accusations, unfounded suspicions, and widespread stigma toward the area, its products, and the evacuees themselves. As a result, as of August 2017, there were still around 58,000 evacuees [6].

Since February 2012, Fukushima Medical University (FMU) has been conducting major population-based mental health surveys (the Mental Health and Lifetime Survey, MHLS) involving approximately 210,000 people who had previously lived in the evacuation area [7]. In these surveys, questionnaires, including a version of the PTSD checklist for specific stressors (PCL-S) [8], were mailed to the targeted population annually. The findings showed that 21.6% of the adults surveyed scored above the cutoff value ($\geq$44) on the PCL-S at 10 months after the disaster, which is almost equal to that of workers after the 9/11 World Trade Center attacks in the

USA [9]. In another recent report on the same population, over 10% of the respondents still showed symptoms of PTSD [10].

In addition to PTSD, the results of the MHLS, which were based on the 6-item Kessler scale (score ≥13), showed that the prevalence of probable depression among adult evacuees was as high as 14.6% in 2012, 11.9% in 2013, and 9.7% in 2014 [11, 12]. Despite this gradual decrease, the scores were still considerably higher than those of the general Japanese population (3%) [13]. In addition, the prevalence of depressive symptoms over the most recent 3 years has been approximately 7% [10], which could indicate a prolonged course of depression. In addition, a 3-year MHLS trajectory analysis revealed that a negative risk perception regarding the genetic effects of exposure to radiation was strongly associated with depressive symptoms [11, 12].

Moreover, public employees working in disaster-affected areas are likely to be considerably exhausted and depressed. In fact, one study [14] revealed that the current prevalence of depression among all workers belonging to two towns in the coastal area of Fukushima was as high as 17.8%. Both the provision of adequate psychiatric interventions and the establishment of an efficient care system for these workers are urgently needed.

6.2.2 Suicide and Related Issues

The high prevalence rates of people at increased risk for depression and PTSD as described above could result in more serious outcomes such as an increase in the suicide rate. Actually, in Fukushima Prefecture, 101 cases of suicide during the 7 years after the Fukushima disaster were officially certified as disaster related by the Japanese Police Agency [15]; this rate is much higher than that reported in other prefectures such as Iwate and Miyagi, which were mainly affected by the tsunami [15]. Furthermore, initially, the standardized suicide mortality ratio (SMR) decreased gradually after the 2011 disaster (108 in 2010, 107 in 2011, 94 in 2012, and 96 in 2013), but then increased to 126 in 2014, exceeding pre-disaster levels (the reference of the SMR is the average suicide rate among the general population in Japan) [16]. This pattern, an increase after a short-term drop, is similar to that noted in a review by Kölves et al. [17]. Furthermore, another study analyzing panel data from Fukushima Prefecture revealed that male and female suicide rates in evacuation areas increased 3 and 4 years after the disaster, respectively, which differs from rates in other areas within the same prefecture [18]. Such a substantial increase in suicide cases in Fukushima Prefecture could conceivably be the result of the effects of the nuclear power plant accident, rather than those of the earthquake or tsunami. In other words, the decrease in community resilience in Fukushima Prefecture might have caused an increase in suicide attempts.

Another psychiatric problem related to suicide and often seen after natural disasters is alcohol abuse [19, 20]. The MHLS showed that the recent prevalence rate of problem drinking according to CAGE (an acronym for "attempts to Cut back on drinking, being Annoyed at criticisms about drinking, feeling Guilty about drinking, and using alcohol as an Eye opener") scores remained relatively high in both males (17.1% in 2017) and females (9.2% in 2017) [11, 12]. In spite of the lack of CAGE data from the general population in Japan, these findings suggest that primary prevention strategies need to be prioritized for people at increased risk of both alcohol abuse and suicide.

6.3 Social Responses to the Fukushima Accident

6.3.1 *Turmoil in the Initial Phase*

Any natural or man-made accident is associated with complications in the initial stage of the crisis. The Great East Japan Earthquake, which was followed by the subsequent tsunami and nuclear crisis, was no exception. At first, the earthquake caused critical structural damage with implications for seismic safety at several nuclear power plants in Japan, after which, the tsunami crippled the backup electric generators at the Fukushima Daiichi Nuclear Power Plant, disabling its cooling systems and consequently causing a meltdown and large hydrogen explosions at the power station [21]. Subsequent failures in communication resulted in poor crisis management by Tokyo Electric Power Company (TEPCO) and the Japanese government, undermining the credibility of both organizations and causing deep public distrust toward the government, as well as the stigmatization of the entire Fukushima region, multidimensional psychological and social issues among the affected population, and misinformation and contradictory messaging from the mass media, which led to additional widespread groundless rumors across Japan.

The term *fuhyohigai* in Japanese refers to the damage induced by rumors and negative stigmas regarding the people and products affected by the Fukushima disaster, which still prevail within the discourse of the mass media and social networking service [22]. Those who live outside the disaster-stricken area are mostly exposed to harmful and groundless rumor-related media coverage, which ultimately causes *fuhyohigai* both inside and outside of Japan. Initially, the term *fuhyohigai* was mainly used to describe economic damage; however, since the disaster, it has also been used in regard to post-disaster psychological damage [23].

6.3.2 *Social Stigma and Media Coverage*

The mass media are known to play a very important role, such as delivering information to a large number of people simultaneously. Therefore, the mass media usually play one of the major roles in framing and interpreting certain risks, and consequently, directly or indirectly affecting perceptions of different risks among affected populations [24]. In addition, "risks from activities that receive considerable media coverage (e.g., such as accidents and leaks at nuclear power plants) are judged to be greater than risks from activities that receive little attention (e.g., on-the-job accidents)" [25]. After the Chernobyl disaster in 1986, Soviet scientists proved that the way the mass media reported the accident played a major role in the development of post-Chernobyl mental disorders among the affected population in Ukraine, Belarus, and Russia [26]. Fukushima, similar to Chernobyl, was no exception in this regard, and currently faces nearly identical consequences. Just like in case of Chernobyl, the affected Japanese public received substantial amounts of misinformation that often contained inaccurate and contradictory messages [27]. As a result, distrust intensified toward the institutions responsible for the crisis, causing stigma, panic, fear, anxiety, distorted perceptions of radiological risk, and multidimensional social and psychological problems.

Actually, large amounts of evidence can be found in the responses of people who took part in the MHLS. Many affected people described their psychological distress as being related to self-stigmatization, with a focus on both public and self-stigmas. Analyzing their responses, we found two types of stigma-related behaviors among evacuees: "passing" behavior and "covering" behavior, which were devised by the Canadian–American sociologist, Erving Goffman, in 1969. Some people were trying to hide their real-life stories because they experienced the FDNPP accident first-hand. These people were also afraid that the general public might look down on them because of compensation issues and groundless radiation health-related assumptions. One such example is provided below:

> My car license plate was issued in "Iwaki city". So, when someone asks me "are you an evacuee?", my heart starts pounding. I feel that some people envy me because of the possible financial compensation benefits provided by TEPCO; therefore, I cannot answer them honestly. It's quite frustrating… (Anonymous answer from an evacuee in Fukushima Prefecture)

Additionally, one shocking event was broadly reported by the press in November 2016, shedding some light upon the public stigma regarding evacuees in Japan and its psychological impact. For example, a 13-year-old boy was bullied in an elementary school in Yokohama city for being a "nuclear evacuee" from Fukushima. He was called "germ" as a reference to nuclear contamination by his teacher and was forced to pay over ¥1.5 million ($13,500) in extortion fees to his bully classmates [28].

> I am disturbed by media reports about bullying cases related to the nuclear disaster involving evacuees from Fukushima at primary schools. What if my son were in the same situation, what should I do? We have been asking our schoolteachers to keep it a secret that we are evacuees from Fukushima Prefecture. However, I am afraid that our secret will be discovered by students or their parents. (An evacuee living outside Fukushima Prefecture)

A scientific study conducted among people suffering from chronic psychiatric disorders such as schizophrenia showed that self-stigmas caused by public stigmas often made such persons more anxious and unstable, and could even reduce self-efficacy and self-esteem [29]. In Fukushima, we have attempted to apply the Stage Model of Self-Stigma proposed by Corrigan et al. [30] to evacuees (Fig. 6.1). Similar to people with chronic psychiatric disorders, evacuees who are exposed to public stigmas as described above will also most likely experience self-stigmas, and thus more likely to have decreased self-efficacy and suffer from depression and PTSD [31].

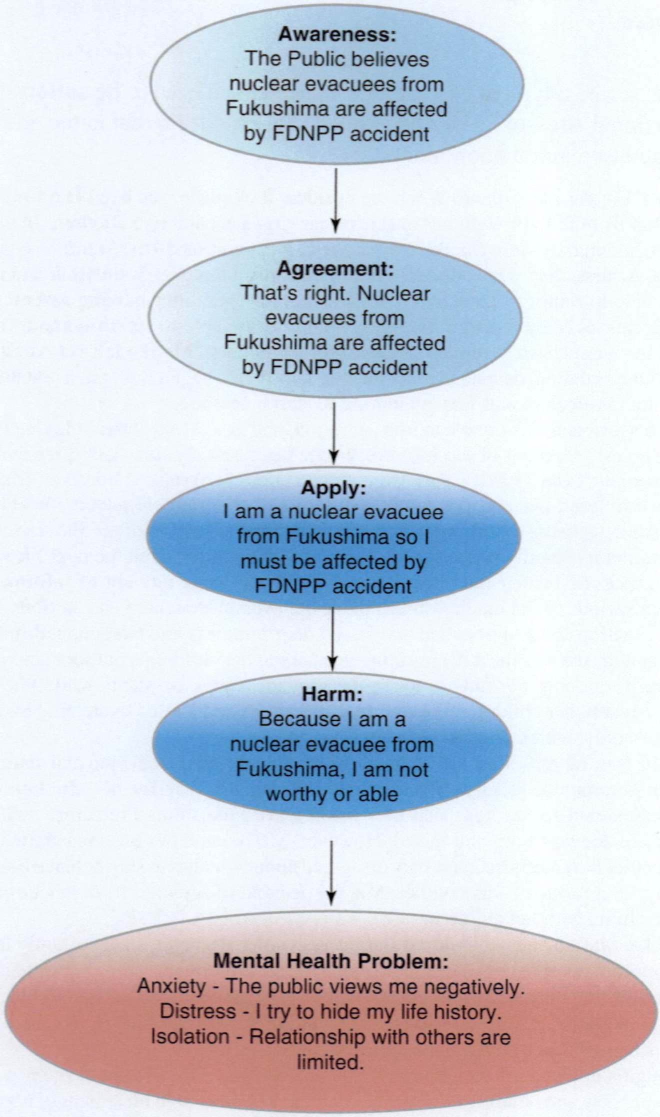

Fig. 6.1 The Stage Model of Self-Stigma among nuclear evacuees from Fukushima. This figure is newly made based on "The Stage Model of Self-Stigma" by Corrigan and Rao [30]. If evacuees from Fukushima are aware of public stigma about their condition ("awareness"), then they may agree that these negative public ideas are true about the group ("agreement"). Subsequently, they concur that these stereotypes apply to themselves ("application"). This may lead to significant decreases in self-esteem and self-efficacy ("harm"), resulting in mental health problems such as depression, psychological distress, and isolation

6.3.3 Case

Below we describe a typical case of an evacuee considered to be suffering from the multidimensional stressors described above (the evacuee's real name and data have been modified to maintain confidentiality).

> Mariko is a 35-year-old woman. When the accident took place, she lived in a town located about 20 km from FDNPP with her husband (car engineer) and two children. Immediately after the accident, they were forced to evacuate in great confusion to several places, including Tokyo. At first, they were confident that it was just a temporary situation and that they would be able to return to their home very soon. However, after moving several times to different locations, they realized that they would not be able to return home as they had expected, but would have to remain evacuees for a long time. Mariko felt very disappointed because of the situation, despite the fact that her family had begun living in a tentative house in Fukushima Prefecture and was attempting to start a new life.
>
> Mariko's husband was unable to find a new job, and as a result, started playing Pachinko (Japan's biggest gambling obsession) every day. Because they had started receiving fixed financial support from TEPCO, they were able to avoid bankruptcy; however, Mariko had great concern about her and her family's future and the possible adverse health effects resulting from exposure to radiation, especially in terms of her children. She started checking the Internet repeatedly every day to search for information about the possible effects of radiation exposure. Rather than becoming relieved, the large amount of information she found from various social media sources made her even more scared and anxious, as much of the information she had obtained was based on groundless and exaggerated rumors.
>
> In particular, she was most deeply concerned about her children's outdoor activities. She soon began forbidding her kids to touch the ground, leaves, or plants while playing outdoors. As a result, her children preferred to stay indoors and started overeating and playing video games only, which ultimately led to obesity.
>
> Luckily, four months after the accident, her husband found a new job and started working for an automobile factory. However, considering her fear for her children's health, Mariko maintained strong hope that they could leave Fukushima Prefecture and move to Osaka, where she was born and raised. However, Mariko and her husband started arguing with each other frequently because they disagreed about whether to stay or leave Fukushima. Eventually, as a result of this conflict, Mariko decided to separate from her husband and leave Fukushima with her children.
>
> In Osaka, Mariko and her children started a new life. However, she constantly felt afraid and anxious that other people might hold some negative feelings against her or her children due to the fact that they had experienced the FDNPP accident, so she tried to hide her real life story. She was also very worried that her children might be bullied at school by their classmates because they were Fukushima evacuees.
>
> Although she faced a lot of troubles, she still hesitated to talk about them with other people in her new surroundings in Osaka. Gradually, she became increasingly nervous and started having sleeping difficulties for three or more days per week. Additionally, her constant negativity about her and her family's hopeless future was taking its toll. She started blaming herself for everything: "I could not protect my children or my husband… I have caused a lot of trouble to everyone…Every day, I feel that I am worthless, that I am ugly". Finally, she started thinking about suicide more frequently and even making actual plans for an attempt. Eventually, Mariko's husband became worried about her unusual and suspicious behavior, so he came to Osaka to meet her and convinced her to visit a psychiatric clinic. Luckily, Mariko started psychiatric treatment about 8 months after the accident.

Mariko was forced to evacuate without any physical and mental preparation for the natural disaster followed by the nuclear crisis. All she wanted was to be able to settle down in a new and safer place. The large amount of information regarding the possible adverse effects of radiation exposure on health only served to confuse her, making it impossible to clearly distinguish between truth and falsehoods and thereby decide on future plans. In particular, she was very concerned about the possible negative consequences of radiation exposure to her children; this resulted in serious conflict with her husband and eventually a marital crisis. Her guilty feelings about her children and husband induced severe depressive symptoms; however, luckily, she decided to follow her husband's recommendation and undertake psychiatric treatment. This case is typical of the psychological distress experienced by many evacuees, especially young mothers, after a disaster. Unfortunately, unlike in this case, because of the widespread prejudice common toward psychiatric disorders and their treatment, in reality, only limited numbers of evacuees actually visited a psychologist or psychiatric clinic after the disaster.

6.4 Currently Available Care Resources in Fukushima and Remaining Challenges

Japan is affected by numerous natural disasters, including regular earthquakes, tsunamis, and typhoons, so physical and psychological suffering after a disaster is not uncommon. Despite causing traumatic experiences, such natural events seem to enhance community resilience and in many cases, strengthen the bonds between residents. In recent decades, knowledge about PTSD has increased, and recently, many people affected by different types of disasters in Japan have had the opportunity to receive quality physical and psychological health care, treatment, and support.

In fact, after the Great East Japan Earthquake and nuclear accident, many different organizations, including non-governmental organizations, have emerged to provide psychological support and health care to affected residents and evacuees. In terms of recent activities, two major new facilities were established after the disaster: the Radiation Medical Science Center (RMSC) at FMU and the Fukushima Center for Disaster Mental Health (FCDMH). The authors of this chapter have been deeply engaged in the activities of both institutions.

6.4.1 Radiation Medical Science Center (RMSC)

The RMSC at FMU was established in 2012 with the objective of carrying out the "Fukushima Health Management Survey," which consists of a basic survey that aims to estimate radiation dose and four additional detailed surveys, namely a thyroid ultrasound examination, a comprehensive health check, the MHLS, and a pregnancy and birth survey [7].

Specifically, the MHLS aims to identify people at risk of a number of psychiatric disorders, such as depression and PTSD, as well as secondary and lifestyle-related issues, such as a lack of physical exercise, alcohol abuse, smoking habits, and obesity. The MHLS uses several different questionnaires which are mailed annually to about 210,000 people living in the evacuation zone across Fukushima Prefecture. After identifying people who have some psychological and/or lifestyle-related problems, brief interventions are usually provided through telephone consultations. In some cases, when a telephone intervention is not possible or the respondent experiences relatively minor challenges, leaflets containing material that provides information about mental health issues among other various problems are sent to the recipients. Thanks to the efforts of both institutions, since 2012, about 3000–5000 people every year have received psychological support [32].

Some important statistical data obtained via the MHLS have already been reported in this article; therefore, we would now like to describe the actual telephone interventions provided, when necessary, after the MHLS. For efficient interventions, the RMSC set up a professional support team composed of 15 counselors with extensive field experience in clinical psychology, social work, and nursing care. During the telephone interventions, the respondents, who are identified as being at increased risk for mental health problems, are typically asked about their recent health condition and related habits. In many cases, the counselors provide useful advice and suggestions to ease the troubles or stress experienced by the respondents. If necessary, the counselors can also refer the respondents to appropriate experts or health care facilities (e.g., psychiatric clinics, local health centers). On average, the telephone counseling sessions last 10–30 min each. In addition, the time needed to contact each individual identified as being in need of intervention takes around 3–4 months [32].

This unique telephone counseling intervention has several limitations. First, the MHLS currently has only one professional team consisting of 15 people who are adequately skilled to conduct direct visit services and/or face-to-face interviews. Considering the fact that there are still around 58,000 remaining evacuees across and outside Fukushima Prefecture [6], and that over 10,000 people are still estimated to be at risk of mental health disorders in the evacuation area, telephone counseling seems to be the best of only a few feasible ways to provide support. The current RMSC team is attempting to establish a network across different health care facilities, such as psychiatric clinics and local health care centers, in order to disseminate necessary information and reach as wide an audience as possible so that existing staff limitations can be overcome and health care can be provided for all evacuees in need.

6.4.2 Fukushima Center for Disaster Mental Health (FCDMH)

The FCDMH was also established in 2012, and like the MHLS, is fully funded by the Japanese government. The major role of the FCDMH is to provide mental health care support to affected people living in Fukushima Prefecture. Pre-disaster care facilities across Fukushima Prefecture were not able to provide adequate support, mainly because of the shortage of human resources. Currently, the FCDMH employs about 40 mental health care professionals, including clinical psychologists, social workers, nurses, and other mental health specialists. To cover the broad area affected by the nuclear accident, the FCDMH currently has branches in the following five locations: Fukushima, Koriyama, Minamisōma, Aizuwakamatsu, and Iwaki [33].

The FCDMH, in cooperation with local residences, is attempting to provide a wide range of activities, including various types of interventions such as outreach services (home visiting services), psychological education, and relaxation methodology. Based on the establishment of the FCDMH, the initial situation in the affected area was extremely complicated and confusing. It was not easy for many of the FCDMH staff to establish a good working relationship with evacuees, and the preexisting health care facilities in the region were lacking because no one had any experience in how to deal with people suffering from the anxieties caused by a nuclear crisis. The professionals at the FCDMH had to learn gradually through trial and error. However, thanks to their continuous efforts, they succeeded in convincing those at the preexisting facilities and other stakeholders in the region to recognize and acknowledge the FCDMH as one of the most useful and important care resources currently available in the region to provide the health care and support necessary for evacuees [33].

Despite the fact that the FCDMH is conducting different types of activities, as mentioned above, the most valuable activities have been the outreach and visiting services. On average, between 4000 and 5000 affected people annually are directly visited by the FCDMH staff for support in cooperation with municipal governments. Urgent crisis interventions are typically provided on a priority basis when people at risk of serious mental health problems and/or suicide are identified. These professionals usually attempt to conduct risk assessments, share information with other care facilities, and recommend that the people at risk visit a nearby psychiatric clinic. In recent years, in addition to the prior focus on evacuees, the professional staff at the FCDMH have also started to provide mental health care support to many public employees working at different municipal offices who face burnout owing to the challenges related to the recovery activities in the remaining disaster-affected areas.

While these efforts have gradually produced good results, new and more difficult and diverse challenges and demands that seem to be beyond the capabilities of local preexisting health care facilities are being encountered. Stressful long-term assignments at the FCDMH have gradually exhausted some of the staff, and as a result, many professionals decide to leave the FCDMH after several years. Therefore, to adequately address the mental health care needs of affected populations, it is important for both the government and policy makers alike to realize that specialized mental health care facilities are indispensable after a major disaster, such as a nuclear accident, and should be operated for a long time (a minimum of 10 years). Local staff who carry the bulk of the burden in providing actual mental health care and support also require continuous support from national and local governments, who should ensure a stable working environment for workers, continuous training, and adequate financial remuneration to maintain the required number, as well as the morale and continuous interest, of professional staff in the mental health field.

References

1. Yabe H, Suzuki Y, Mashiko H, Nakayama Y, Hisata M, Niwa S. Psychological distress after the Great East Japan Earthquake and Fukushima Daiichi Nuclear Power Plant accident: results of a mental health and lifestyle survey through the Fukushima health management survey in FY2011 and FY2012. Fukushima J Med Sci. 2014;60:57–67.
2. Yoshizawa N. Fukushima reconstruction: current status and radiation health risks. 2017. https://www.mri.co.jp/opinion/column/trend/trend_20171226.html. Accessed 16 Aug 2018.
3. Tanigawa K, Lochard J, Abdel-Wahab M, Crick MJ. Roles and activities of international organizations after the Fukushima accident. Asia Pac J Public Health. 2017;29:90S–8S.
4. Kimura HA. Radiation brain moms and citizen science. The gender politics of food contamination after Fukushima. Durham: Duke University Press; 2016.
5. Sato A, Lyamzina Y. Diversity of concerns in recovery after a nuclear accident: a perspective from Fukushima. Int J Environ Res Public Health. 2018;15(2):350E. https://doi.org/10.3390/ijerph15020350.
6. Fukushima Prefectural Government Steps for revitalization in Fukushima. 2017. http://www.pref.fukushima.lg.jp/uploaded/attachment/233344.pdf. Accessed 16 Aug 2018.
7. Yasumura S, Hosoya M, Yamashita S, Kamiya K, Abe M, Akashi M. Study protocol for the Fukushima health management survey. J Epidemiol. 2012;22:375–83.
8. Iwasa H, Suzuki Y, Shiga T, Maeda M, Yabe H, Yasumura S. Psychometric evaluation of the Japanese version of the posttraumatic stress disorder checklist in community dwellers following the Fukushima Daiichi nuclear power plant incident: the Fukushima health management survey. SAGE Open. 2016;6(2):1–11.
9. Stellman JM, Smith RP, Katz CL, et al. Enduring mental health morbidity and social function impairment in world trade center rescue, recovery, and cleanup workers: the psychological dimension of an environmental health disaster. Environ Health Perspect. 2008;116:1248–53.
10. Mental Health and Lifestyle Survey for the Fukushima Health Management Survey 2018. http://fmu-global.jp/download/results-of-mental-health-and-lifestyle-survey-for-fy-2016-summary-of-support/?wpdmdl=4420. 2018. Accessed 16 Aug 2018.
11. Oe M, Fujii S, Maeda M, Nagai M, Harigane M, Miura I, et al. Three-year trend survey of psychological distress, posttraumatic stress, and problem drinking among residents in the evacuation zone after the Fukushima Daiichi nuclear power plant accident: the Fukushima health management survey. Psychiatry Clin Neurosci. 2016;70:245–52.
12. Oe M, Maeda M, Nagai M, Yasumura S, Yabe H, Suzuki Y, et al. Predictors of severe psychological distress trajectory after nuclear disaster: evidence from the Fukushima health management survey. BMJ Open. 2016;6:e013400. https://doi.org/10.1136/bmjopen-2016-013400.
13. Kawakami N. National survey of mental health measured by K6 and factors affecting mental health status in Research on Applied Use of Statistics and Information. Health Labour Sciences Research Grant (in Japanese). 2007.
14. Maeda M, Ueda Y, Nagai M, Fujii S, Oe M. Diagnostic interview study of the prevalence of depression among public employees engaged in long-term relief work in Fukushima. Psychiatry Clin Neurosci. 2016;70:413–20.
15. Japanese Cabinet Office. Statistical data regarding suicide (in Japanese). 2017. http://www8.cao.go.jp/jisatsutaisaku/toukei/pdf/saishin_shinsai.pdf. Accessed 16 Aug 2018.
16. Ohto H, Maeda M, Yabe H, Yasumura S, Bromet EE. Suicide rates in the aftermath of the 2011 earthquake in Japan. Lancet. 2015;38:1727.
17. Kõlves K, Kõlves KE, De Leo D. Natural disasters and suicidal behaviours: a systematic literature review. J Affect Disord. 2013;146:1–14.
18. Orui M, Suzuki Y, Maeda M, Yasumura S. Suicide rates in evacuation areas after the Fukushima Daiichi nuclear disaster: a 5-year follow-up study in Fukushima Prefecture. Crisis. 2018;5:1–11. https://doi.org/10.1027/0227-5910/a000509. [Epub ahead of print].
19. Beaudoin CE. Hurricane Katrina: addictive behavior trends and predictors. Public Health Rep. 2011;126(3):400–9.

20. Cerdá M, Tracy M, Galea S. A prospective population based study of changes in alcohol use and binge drinking after a mass traumatic event. Drug Alcohol Depend. 2011;115(1-2):1–8.
21. Kingston J. In: Nissan Institute, editor. Natural disaster and nuclear crisis in Japan: response and recovery after Japan's 3/11, Routledge Japanese studies series. London: Routledge; 2012. ISBN-13: 978-0415698566.
22. Hirakawa H, Shirabe M. Rhetorical marginalization of science and democracy. In: Fujigaki Y, editor. Lessons from Fukushima: Japanese case studies on science, technology and society. New York: Springer; 2015. p. 69–71.
23. Sekiya N. What is fuhyohigai? Fukushima J Med Sci. 2011;57:93–9.
24. Vyncke B, Perko T, Van Gorp B. Information sources as explanatory variables for the Belgian health related risk perception of the Fukushima nuclear accident. Risk Anal. 2017;37(3):570–82.
25. Covello V, Sandman P. Risk communication: evolution and revolution. In: Wolbarst A, editor. Solutions to an environment in peril. Baltimore: John Hopkins University Press; 2001. p. 164–78.
26. Lazarev V. Psychological-and-emotional problems of Chernobyl: role of mass media in the development of post Chernobyl mental disorders in population: critical-and-analytical review. In: Information science and science of science: proceedings of the 3rd international (5th Tambov) conference. Tambov; 1999. pp. 54–57.
27. Tomkiv Y, Perko T, Oughton DH, et al. How did media present the radiation risks after the Fukushima accident: a content analysis of newspapers in Europe? J Radiol Prot. 2016;36(2):S64–81.
28. The Japan Times. 2017. https://www.japantimes.co.jp/news/2017/06/01/national/social-issues/yokohama-mayor-apologizes-boy-bullied-fukushimanuclear-evacuee/#.XQmg1Yj7Rdg. Accessed 19 June 2019.
29. Corrigan PW, Watson AC, Barr L. The self-stigma of mental illness: implication for self-esteem and self-efficacy. J Soc Clin Psychol. 2006;25:875–84.
30. Corrigan PW, Rao D. On the self-stigma of mental illness: stages, disclosure, and strategies for change. Can J Psychiatr. 2012;57(8):464–9.
31. Maeda M, Oe M. The Great East Japan Earthquake: tsunami and nuclear disaster. In: Cherry KE, editor. Traumatic stress and long-term recovery coping with disasters and other negative life events. New York: Springer; 2015. p. 71–90.
32. Kashiwazaki Y, Maeda M, Yagi A, Fujii S, Takahashi N, Yabe H, et al. Effectiveness of telephone-based intervention for people living in Fukushima disaster area/Fukushima health management survey. Seishinigaku. 2016;58:433–42 (in Japanese).
33. Maeda M, Ueda Y, Hiruta G. What is the role of Fukushima Center for Disaster Mental Health? Jpn J Trauma Stress. 2014;12:5–12 (in Japanese).

Chapter 7
Impact of Evacuation on Lifestyle Activity and Metabolic Status Following the Fukushima Daiichi Nuclear Power Plant Accident: Preliminary Findings

Takashi Eto, Yun-shan Chung, Daniel K. Ebner, Kouji H. Harada, Jinro Ishizuka, Keiko Igari, and Akio Koizumi

Abstract On March 11, 2011, Kawauchi Village, in Fukushima Prefecture, was struck by both the Tohoku earthquake and the Fukushima Daiichi Nuclear Power Plant (FDNPP) accident, with subsequent evacuation. One year later, in April 2012, villagers began to return. In the previous studies, increases in metabolic diseases were noted; here, we compared the prevalence of metabolically unhealthy conditions among evacuee and non-evacuee populations located nearby, namely Miharu and Ono Towns. This is a retrospective study using health examination data comparing health and health changes prior to the 2011 disaster (2008–2010: baseline) to the post-disaster period (2012–2015). 3451 residents who attended annual health checkups both between 2008 and 2010 and after 2011 made up the study population (599 evacuees and 2852 non-evacuees, collectively "baseline"). Disease states examined in this study included diabetes mellitus (DM) and borderline DM, poly-

T. Eto · A. Koizumi (✉)
Department of Health Environmental Sciences, Kyoto University Graduate School of Medicine, Kyoto, Japan

Public Interest Corporation Kyoto Hokenkai Nakagyouku, Kyoto, Japan
e-mail: koizumi@kyoto-hokenkai.or.jp

Y.-s. Chung · K. H. Harada
Department of Health Environmental Sciences, Kyoto University Graduate School of Medicine, Kyoto, Japan

D. K. Ebner
Harvard TH Chan School of Public Health, Boston, MA, USA

J. Ishizuka
Medical Corporation Ishizuka Clinics, Ono, Fukushima, Japan

K. Igari
Yufune Health Care Center, Kawauchi Village Office, Fukushima, Japan

G. Steinhauser et al. (eds.), *Nuclear Emergencies*,
Current Topics in Environmental Health and Preventive Medicine,
https://doi.org/10.1007/978-981-13-8327-4_7

cythemia, hypertension, obesity, as well as exercise and smoking habits. We confirmed increases in the prevalence of obesity, diabetes, and polycythemia, which continue even 4 years after the FDNPP accident in the evacuee population, when compared to age- and gender-matched neighborhood populations. On qualitative questionnaire, evacuees were noted to have a higher rate of habitual exercise. The reasons for this shift in population health remain under investigation, but may involve a shift from active to sedentary lifestyle in the evacuee population that cannot be compensated for by increased exercise habit.

Keywords Evacuation · Lifestyle · Diabetes · Metabolic · FDNPP accident

7.1 Introduction

On March 11, 2011, Kawauchi Village, in Fukushima Prefecture, was stuck by both the Tohoku earthquake and the subsequent Fukushima Daiichi Nuclear Power Plant (FDNPP) accident (Fig. 7.1). The village lies 20 km west of the FDNPP and the entire village population evacuated to a nearby evacuation center in Koriyama City on March 16, 2011. A year later, in April 2012, villagers began to return as environmental radiation levels were confirmed to be safe. Although radiation contamination

Fig. 7.1 Map of the study area: Kawauchi Village, Ono Town, Miharu Town, and Fukushima Daiichi Nuclear Power Plant (FDNPP)

in Kawauchi village was limited, the proximity to the power plant led to evacuation, resulting in increases in unemployment, loss of living area, and disintegration of families and other social support systems, posing innumerable challenges to the returning population.

Several studies describing the acute effects of up to 3 months of evacuation on health after the Great Hanshin earthquake in 1995 have been published [1, 2] and have been used for establishing preventive measures in Japan. However, there are scant data for the long-term effects of such disasters and evacuations on human health over the long term. Recently, several studies have reported rapid increases in the prevalence of metabolically unhealthy conditions, such as obesity, diabetes, chronic kidney disease, and polycythemia, in the coastal area of Fukushima where many residents were evacuated [3–10]. Those observations are, however, mostly limited to the evacuee population and cannot eliminate the effects of population aging. Thus, it remains unknown whether increases in the prevalence of metabolically unhealthy conditions are associated with evacuation, population aging, or both. In the present study, we compared the prevalence of metabolic unhealthy conditions among evacuee and non-evacuee populations in geographically close neighborhood communities, Miharu Town and Ono Town. We designed this study so as to control for the effects of population aging, and allow for evaluation of metabolic changes in the evacuated population, as well as delineate the magnitude of the long-term effects acting on this population.

7.2 Materials and Methods

7.2.1 Study Population and Methods

Kawauchi Village lies in the Futaba District of Fukushima Prefecture, and was generally undamaged by the 2011 earthquake and tsunami. As it was in the 20-km evacuation zone, it was nonetheless forced to evacuate (villagers are hereafter referred to as "evacuees"). The nearby villages of Miharu and Ono (Tamura-gun, Fukushima Prefecture) lay outside the zone, and did not evacuate ("non-evacuees").

This is a retrospective study using health examination data comparing health and health changes prior to the 2011 disaster (2008–2010) to the post-disaster period (2012–2015). The examination data used is compiled each year by the Health Welfare divisions of three villages: Kawauchi Village, Miharu Town, and Ono Town, in accordance with National Law/Program that conducts evaluations of seniors. This data was anonymized by the Health Welfare divisions, with IDs provided allowing for tracking of individual residents between multiple years. The National Insurance System in Japan provides elderly individuals (ages 65 and over) with a yearly health checkup. At the time of analysis, i.e., July 2017, data for 7058 residents was available. Data from 2011 was excluded due to low rates of examination, as well as variable examination timing owing to the disaster. 3451 residents who attended annual health checkups between 2008 and 2010 and after 2011 were identified, and constitute the study population (599 evacuees and 2852 non-evacuees, collectively "baseline"). If individuals attended annual health checkups several times between 2008 and 2010, the most recent data pre-disaster data was selected.

Disease states examined in this study are defined as: diabetes is classified into "diabetes (DM)" and "borderline DM," with a fasting blood glucose greater or equal to 126 mg/dL or HbA1c greater or equal to 6.5% for diabetes, and fasting blood glucose levels greater than or equal to 110 mg/dL or HbA1c greater than or equal to 6.0% for borderline DM. Polycythemia was defined by hemoglobin (Hb) levels exceeding the upper 25% limit by gender at baseline in each area: 15.5 g/dL for males and 13.7 g/dL for females in Kawauchi; 15.4 g/dL and 13.9 g/dL in Miharu; and 15.2 g/dL for males and 13.6 g/dL for females in Ono. Hypertension was defined as greater than or equal to 140 mmHg for systolic blood pressure or greater than or equal to 90 mmHg for diastolic blood pressure. Obesity is defined by body mass index (BMI) greater than or equal to 25. Exercise and\or smoking habits were evaluated qualitatively by questionnaire, asking "yes" or "no" to the question "Did you have 30-min-sweat-generating exercise at least two times per week for the last year?" or "Are you current smoker or not," respectively.

The use of patient healthcare data was approved by the mayors of Kawauchi, Miharu, and Ono, and anonymized prior to receipt by the analysis team. This study was approved by the IRB/Ethics Committee of Kyoto University (R0869).

7.2.2 *Statistical Analysis*

Regional comparison of evacuee vs non-evacuee groups was conducted using the Student's t-test for continuous values and the Chi-square test for categorical variables. For comparison between the pre- and post-disaster time periods, ANOVA and chi-square were used for continuous and categorical variables, respectively. Significance was set at a p level less than 0.05. Statistica (Dell Software, CA, USA) was used for all statistical analysis.

7.3 Results

Table 7.1 described the demographic characteristics for the study population. At baseline, 599 evacuees and 2852 non-evacuees joined this study. Mean age and female ratio were homogeneous between evacuee population and non-evacuee population in an observation period from baseline to 2015. During the observation period, the ratio of male to female did not differ between the two groups.

Obesity was significantly and consistently more prevalent in evacuees (40.5–41.3%) than in non-evacuees (31.4–30.1%) after 2012, while at baseline its preva-

Table 7.1 The prevalence of obesity, hypertension, DM, borderline DM, and polycythemia among evacuees and non-evacuees

		Baseline	2012	2013	2014	2015	*ANOVA*
Examinees (numbers)	Evacuee	599	466	426	405	430	(–)
	Non-evacuee	2852	2348	2172	1878	2009	(–)
Age (years)	Evacuee	66.7 ± 10.4	69.7 ± 10.6	70.2 ± 10.0	70.6 ± 9.7	70.9 ± 9.7	<0.0001
	Non-evacuee	66.0 ± 9.3	68.9 ± 8.9	69.3 ± 8.9	70.4 ± 8.9	70.6 ± 8.6	<0.0001
Female ratio (%)	Evacuee	58.3	57.3	58.2	59.5	57.4	0.96
	Non-evacuee	55.2	55.5	56.1	54.5	54.4	0.85
Obesity (%)	Evacuee	34.7	40.5[a]	37.1[a]	40.4[a]	41.3[a]	0.2
	Non-evacuee	30.5	31.4	29.9	30.4	30.1	0.85
Hypertension (%)	Evacuee	63.2[a]	66.1[a]	66.0[a]	61.5	66.8[a]	0.45
	Non-evacuee	54	56.3	56.5	54.8	57.6	0.11
Diabetes (%)	Evacuee	9.7	13.3	16.2[a]	15.4[a]	17.5[a]	<0.0001
	Non-evacuee	8.7	10.5	11.3	11.7	13.1	<0.0001
Borderline DM (%)	Evacuee	7.2	11.1	16.8	15.4	17.0	<0.0001
	Non-evacuee	11.3[a]	10.8	13.6	14.5	15.9	<0.0001
Polycythemia (%)	Evacuee	23.7	30.2[a]	35.7[a]	36.6	36.4[a]	<0.0001
	Non-evacuee	22.8	23.2	20.4	31.3	22.9	<0.0001
Exercise[b] (%)	Evacuee	33.8	42.6[a]	45.9[a]	44.7[a]	41.7[a]	0.001
	Non-evacuee	31.8	32.8	35	32.3	32.7	0.34
Smoking[c] (%)	Evacuee	16.7	9.1	10.3	11.1	11.4	0.17
	Non-evacuee	19.4	11.8	12.3	10.5	12.2	0.001

[a]Significantly larger when compared between evacuees and non-evacuees ($p < 0.05$)
[b]Percentage of those who were doing 30 min of sweat-generating exercise at least two times per week for at least a year
[c]Percentage of current smokers

lence was not different between the two groups (34.7% vs 30.5%). Hypertension was significantly more prevalent in evacuees (63.2%) than in non-evacuees (54%) at baseline and in the following periods (66.1–66.8% vs 56.3% vs 57.6%, respectively).

Although the prevalence of diabetes showed an increasing trend in both evacuees and non-evacuees, diabetic prevalence in evacuees had increased significantly from 9.7% to 17.5% and in non-evacuees from 8.7% to 13.1% during the observation period. Although after 2014 difference in the prevalence of borderline DM between the evacuees and the non-evacuees became smaller than in 2013, it should be noted that the prevalence at baseline was higher in non-evacuees than evacuees and the prevalence of diabetes was still significantly higher in evacuees than in non-evacuees. The prevalence of polycythemia was steadily increasing in evacuees by 1.5 (23.7% vs 36.4%) while it did not change in non-evacuees (22.8% vs 22.9%).

Exercise habits dramatically increased in evacuees after the FDNPP accident (33.8% vs 41.7%), while it did not in non-evacuees (31.8% vs 32.7%). More than 40% of the evacuees answered that they had or had developed an exercise habit after 2011. The prevalence of smoking has been decreasing in both populations and there was no difference between the two populations during the observation period.

7.4 Discussion

In the present study, increases in the prevalence of obesity, diabetes, borderline DM, and polycythemia, which had been reported in evacuees in Fukushima [10] in the earlier periods, were confirmed to continue even 4 years after the FDNPP accident in the evacuee population. As expected, however, we also found persistent increases in diabetes and borderline DM in the non-evacuee population. It has been uncertain what portions of those changes can be attributable to evacuation or population aging or both [7]. In the present study, we are able to show that, in comparison with an unevacuated control population, the prevalence of diabetes and borderline DM in the evacuee population was overwhelmingly greater. This result is noted with no observed difference in mean age or gender ratio between the two populations, with the rates increasing in both populations over the time frame of the study, as may be expected in an aging population. Furthermore, we could demonstrate that the increased prevalence of obesity and of polycythemia was only observed in the evacuee population. Collectively, this study reveals significant increases in the prevalence of diabetes above potential confounding effects such as population aging, and identified evacuation-specific increases in obesity and polycythemia.

Psychological stress has been reported to be the most suspected aggravating factor for diabetes, borderline DM, and obesity after the FDNPP accident [11, 12]. The evacuees have been reported to be exposed to heavy psychological stress, which is known to aggravate diabetes [13]. The evacuees experienced disaster-induced changes in socioeconomic status: diminished privacy, limited food availability, threat of unemployment, reduced income, damages of property, contamination of rice fields, and health concerns [14].

The consequences of psychological stress on evacuee health deserve further evaluation. These may include poor dietary control, limited physical activity, as well as increasingly sedentary lifestyle. It should be addressed, however, that the self-reported questionnaire administered in this study indicated that evacuees enhanced their healthy lifestyle by increasing their exercise habits in comparison with the non-evacuee population, suggesting that the worsening of health outcomes is occurring despite an increase in exercise within the evacuee population. This may mask other negative consequences of the disaster on health, which cannot be captured here.

We speculate that these negative factors are associated with disruption of the traditional-three-generation family structures and house-labor styles prevalent in this region prior to the disaster: generations who previously supported the household were forced to leave Kawauchi village because they lost their jobs due to the shutdown of FDNPP, not returning following the study period. Though these permanent evacuees were not included in the study population, their friends and relatives were: judging from the mean age, most of the current study population is retired, the older members of the three-generation families, who would normally play a supporting role in the traditional house structure, including childcare and farming. The removal of the two younger generations from this structure would likely lead to changes in dietary habits, while also reducing the amount of physical exercise and labor expended by the population. As such, the medical impact seen in the evacuee population may stem from a diminishment of daily activity and communal involvement beyond what an improvement in exercise habits can capture. Future studies will evaluate this by evaluating the currently identified metabolic outcomes versus changes in daily activity and diet between populations within the region.

A previous study [10] reported that the prevalence of polycythemia significantly increased among evacuees compared with non-evacuees. This increase could be detectable even when stratified by smoking and obesity prevalence. Although it was not clear how evacuation led to polycythemia biologically, these authors assumed that mental stress might be a reason. The authors assumed a logical chain, in which mental stress increased hypertension, which is a well-known cause of polycythemia [10]. In this study, we also confirmed that smoking status, which is a well-known risk factor for polycythemia [15, 16], in the two populations is not different. However, the prevalence of hypertension did not change in either evacuee population or non-evacuee population from baseline to 2015 in the current study (Table 7.1). Thus we cannot attribute the increase in the prevalence of polycythemia to hypertension. Alternatively, we may assume that a sedentary lifestyle is a major risk factor for polycythemia. We previously investigated Hb levels among farmers in a nationwide survey including Fukushima in the 1980s [17], finding Hb levels changed with workload, increasing in the off-season but decreasing in the active farming season [17]. Extending this, it may be that an increasingly sedentary life secondary to the psychosocial factors elaborated above is also contributing to an increase in polycythemia rates in Kawauchi. A biological mechanism for this remains unknown. From the viewpoint of disease prevention, an increased Hb level may require intervention due to a connection with incidence of cardiovascular and cerebrovascular diseases [12].

We are now considering that evacuation shifted active lifestyle to sedentary lifestyle in evacuee population. Several studies have reported the usefulness of revitalization of social capital to prevent metabolic diseases [18]. Social capital may likely be a supportive role in place of the three-generation family. In the evacuee population, we propose to reconstruct social capital in the evacuee population in parallel with individual intervention.

Acknowledgments We are grateful to residents in Kawauchi Village, Miharu Town, and Ono Town.

References

1. Saito K, Kim JI, Maekawa K, Ikeda Y, et al. The great Hanshin-Awaji earthquake aggravates blood pressure control in treated hypertensive patients. Am J Hypertens. 1997;10:217–21.
2. Kirizuka K, Nishizaki H, Kohriyama K, Nukata O, et al. Influences of The Great Hanshin-Awaji Earthquake on glycemic control in diabetic patients. Diabetes Res Clin Pract. 1997;36:193–6.
3. Tsubokura M, Takita M, Matsumura T, Hara K, et al. Changes in metabolic profiles after the Great East Japan Earthquake: a retrospective observational study. BMC Public Health. 2013;13:267. https://doi.org/10.1186/1471-2458-13-267.
4. Satoh H, Ohira T, Hosoya M, et al. Evacuation after the Fukushima Daiichi nuclear power plant accident is a cause of diabetes: results from the Fukushima Health Management Survey. J Diabetes Res. 2015;2015:623790.
5. Ohira T, Hosoya M, Yasumura S, et al. Evacuation and risk of hypertension after the Great East Japan Earthquake: the Fukushima Health Management Survey. Hypertension. 2016;68:558–64. https://doi.org/10.1161/HYPERTENSIONAHA.116.07499.
6. Satoh H, Ohira T, Nagai M, et al. Hypo-high-density lipoprotein cholesterolemia caused by evacuation after the Fukushima Daiichi nuclear power plant accident: results from the Fukushima Health Management Survey. Intern Med. 2016;55:1967–76. https://doi.org/10.2169/internalmedicine.55.6030.
7. Ebner DK, Ohsawa M, Igari K, et al. Lifestyle-related diseases following the evacuation after the Fukushima Daiichi nuclear power plant accident: a retrospective study of Kawauchi Village with long-term follow-up. BMJ Open. 2016;6:e011641. https://doi.org/10.1136/bmjopen-2016-011641.
8. Nomura S, Blangiardo M, Tsubokura M, et al. Postnuclear disaster evacuation and chronic health in adults in Fukushima, Japan: a long-term retrospective analysis. BMJ Open. 2016;6:e010080. https://doi.org/10.1136/bmjopen-2015-010080.
9. Hayashi Y, Nagai M, Ohira T, et al. The impact of evacuation on the incidence of chronic kidney disease after the Great East Japan Earthquake: the Fukushima Health Management Survey. Clin Exp Nephrol. 2017;21:995. https://doi.org/10.1007/s10157-017-1395-8.
10. Sakai A, Nakano H, Ohira T. Persistent prevalence of polycythemia among evacuees 4 years after the Great East Japan Earthquake: a follow-up study. Prev Med Rep. 2017;5:251–6. https://doi.org/10.1016/j.pmedr.2017.01.003.
11. Hasegawa A, Ohira T, Maeda M, et al. Emergency responses and health consequences after the Fukushima accident; evacuation and relocation. Clin Oncol. 2016;28:237–44.
12. Ohira T, Nakano H, Nagai M, et al. Changes in cardiovascular risk factors after the Great East Japan Earthquake: a review of the comprehensive health check in the Fukushima Health Management Survey. Asia Pac J Public Health. 2017;29(2S):47S–55S. https://doi.org/10.1177/1010539517695436.
13. Hanson CL, Henggeler SW, Burghen GA. Mode of associations between psychosocial variables and health-outcome measures of adolescents with IDDM. Diabetes Care. 1987;10:752–8.
14. Nagai M, Ohira T, Zhang W, et al. Lifestyle-related factors that explain disaster-induced changes in socioeconomic status and poor subjective health: a cross-sectional study from the Fukushima Health Management Survey. BMC Public Health. 2017;17:340. https://doi.org/10.1186/s12889-017-4247-2.
15. Smith JR, Landaw SA. Smokers' polycythemia. N Engl J Med. 1978;298:8–10.
16. Moore-Gillom JC, Pearson TC. Leading article: smoking, drinking and polycythemia. Br Med J. 1986;292:1617–8.
17. Watanabe T, Ishihara N, Miyasaka M, et al. Hemoglobin levels among Japanese farmers: with special reference to climate and work intensity. Hum Biol. 1986;58:197–208.
18. Hasegawa M, Murakami M, Takebayashi Y, et al. Social capital enhanced disaster preparedness and health consultations after the 2011 Great East Japan Earthquake and nuclear power station accident. Int J Environ Res Public Health. 2018;15:516. https://doi.org/10.3390/ijerph15030516.

Chapter 8
After the Meltdown: Post-Fukushima Environmentalism and a Nuclear Energy Industrial Complex in Japan

Michael C. Dreiling, Tomoyasu Nakamura, Nicholas Lougee, and Yvonne A. Braun

Abstract Since the Kyoto Protocol, and prior to the Fukushima Daiichi meltdown in 2011, a nuclear safety discourse bolstered the industry and supported its promise to meet energy demands while mitigating the risks of global warming. The Fukushima nuclear accident and humanitarian crisis caused by the Tohoku earthquake and tsunami in 2011 called the entire nuclear industry into question. However, the silence among established environmental organizations continued. Based on an extensive study of Japanese environmental organizations, we investigate why Japanese environmental organizations were relatively silent on one of the largest environmental crises in the country's history. We address this question historically and quantitatively, incorporating survey data on a national sample of Japanese environmental organizations. This research also quantitatively operationalizes the effects of board of director composition on the behavior of a subsample of environmental organizations (EOs). The statistical analyses show that environmental groups with government and corporate board members are significantly less likely to publicly denounce nuclear energy following the Fukushima meltdown. The political-organizational embeddedness of these EOs is illustrated with a network heuristic of overlapping industry, government, and environmental organizations, and the implications for civil society and nuclear accident preparedness are discussed.

Keywords Japanese environmentalism · Environmental movements · Disaster politics · Nuclear energy · Network analysis · Civil society · Environmental organizations · Nuclear risk · Fukushima · Japan

---

M. C. Dreiling (✉) · N. Lougee · Y. A. Braun
University of Oregon, Eugene, OR, USA
e-mail: dreiling@uoregon.edu

T. Nakamura
Senshu University, Tokyo, Japan

G. Steinhauser et al. (eds.), *Nuclear Emergencies*,
Current Topics in Environmental Health and Preventive Medicine,
https://doi.org/10.1007/978-981-13-8327-4_8

The Fukushima Daiichi nuclear crisis in Japan in 2011 has presented social scientists with a historical case to assess the efficacy of various institutions in holding accountable powerful energy industries and government agencies captured by industry interests. Ever since US President Eisenhower promised "atoms for peace" in 1953, nuclear energy industries have developed in close proximity to post-WW II US allies, including Japan. Since then, critics have pointed out the ways in which that industry has captured government regulatory agencies and expended enormous capital to support their industry, even at the cost of public safety. Rather than promoting scientifically validated safety or regulatory standards, and public knowledge of the risks and benefits of the industry, critics allege that these agencies have been subordinate to the industry. Referring to Japan, Hasegawa [1, 2] describes an "atomic village," Funabashi [3] references a "nuclear complex," and Kingston [4] elaborates that Japan, which outspends all other countries on nuclear development, "is at the center of the global nuclear-industrial complex." In the face of these conditions, what are civil societies and the larger public to do with knowledge of the varied risks and benefits associated with different national energy priorities? How can civil society groups uphold the promise of democracy and the role of science in public policy by becoming a partner in reasoned public deliberations, in the modern "risk society" [5]?

Risk assessment and preparedness for nuclear emergencies and other crises necessarily require regulatory autonomy and informed public engagement in policy development. How are citizens in a democracy to engage, publicly and privately, the direction of energy policy, nationally and globally? Specifically, what factors help or hinder social movement organizations within civil society in this critical public dialogue concerning the environmental and human costs of different energy priorities and foster clear-eyed preparation for nuclear emergencies in the interest of public safety?

In 2011, geophysicists Noggerath, Geller, and Gusiakov published an article in the *Bulletin of the Atomic Scientists* titled, "Fukushima: The myth of safety, the reality of geoscience." These scientists placed the recent Fukushima disaster within historical perspective, raising key questions about decisions regarding safety and siting of nuclear power plants on Japan's coastline:

> Altogether the historical catalogue counts up to 70 tsunamis generated by submarine earthquakes that have occurred since AD 869 near the eastern Tohoku coast (Iida, 1984; Watanabe, 1998). They include at least six destructive tsunamis… that resulted in run-ups of 25 to 38 meters and thousands of fatalities ([6], p. 39)

In fact, relatively recent incidents had revealed the risks of nuclear power plants in seismically active coastal areas, as well as the broader risks (unrelated to seismic events) of nuclear power. In 1981, an INES Level 2 leak occurred at Tsuruga, resulting in the overexposure of 100 workers during repairs to the facility. A lesser Level 2 incident occurred in the Shika plant in June 1999, resulting in an uncontrolled reaction. Third, the Tokai-mura nuclear accident involved a Level 4 incident at a uranium processing facility in Ibaraki Prefecture in September 1999. Two workers were fatally exposed to neutron radiation and over 100 more were exposed.

Noggerath et al. [6] point to the widespread disbelief about the lack of precautions in coastal Japan after Fukushima:

> The whole world was surprised that precautions against tsunamis were so weak at a nuclear power station located on Japan's coast. How was this design permitted during the station's construction in the 1960s and 1970s, and why were no additional safety measures taken in the time since?

Threats from nuclear energy development have long been contested, with the environmental movement and the nuclear industry challenging the narratives about the relative risks and benefits within public discourse [7]. While the risks of nuclear power for seismically active Japan appeared to be evident, interlocking efforts to deny or minimize these risks, while amplifying the benefits of nuclear power, were deeply entrenched within networks of powerful actors within government, industry, media, and even some environmental organizations. By combining the geophysical risks with a public relations campaign that obscured these facts, a nuclear safety myth [3] functioned as a decades-plus negation of the costs and dangers of the nuclear industry. A nuclear safety myth, like climate denialism, relies on "scientific ambiguity that provides cover for governmental and commercial interests…" ([8], p. 57).

Contemporaneously, nuclear power presents both risks of nuclear disaster and an alternative to the risk of continued fossil fuel dominance in the era of climate change. As climate change sharpens the urgency of alternative energy options, the threats and challenges associated with nuclear power have been softened in policy circles, and nuclear energy recast as a green choice. The contemporary debates about nuclear energy vacillate in the grey area between these two poles, providing a backdrop for the kind of ambiguity Perrow describes above. Within Japan, the historical geophysical risks associated with seismic and tsunami activity compounded the threats associated with the political foreclosures on real debate about the realities of nuclear energy investment in seismically active coastal areas.

The nuclear safety myth was promulgated by industry, government, and media interests who shared overlapping networks of board and executive members who rotated through these institutions over time. Some of these also shared board connections with environmental organizations, which then remained silent in the wake of one of the worst nuclear events in history, the Fukushima nuclear accident in 2011.

Environmental organizations (EOs) as a field have struggled with addressing and contributing to an independent civil society, and avoiding the risks of being captured by political parties or industry interests. Independent environmental organizations are necessary to help guide debate and policy, and to ensure the workings of a robust civil society. A small percentage of EOs that did denounce Japan's nuclear energy development were not captured by the nuclear industry, and our analysis demonstrates that they exhibit high levels of independence, as evidenced in their board composition.

This research explores these issues in the case of environmental movements and the energy sector in Japan. It demonstrates the importance of examining board connections among environmental groups, the government, and industry, and examining what these connections may tell us about their relative independence and autonomy. We look at this in respect to the largest movement concerned with both the risks of climate change, including finding least risk energy policy solutions, and the risks of nuclear disasters, heightened from seismic and tsunami activity in the Japanese islands. We argue that an independent civil society is needed in order to foster robust public discussion about nuclear risks and to implement actions within public and private institutions for preparedness for nuclear emergencies.

8.1 Nuclear Energy and Crisis in Japan

In the wake of the 9.0 Tohoku earthquake and tsunami on March 11, 2011, which tore through the Fukushima Daiichi power plant, the future of nuclear energy in Japan was thrown into serious question.[1] Following the horrendous disaster and nuclear meltdown, large citizen mobilizations escalated from the spring of 2011 to the late summer of 2012. Demonstrations of 20,000–75,000 people were reported several times in Tokyo, with protesters chanting "Sayonara nuclear power."[2] Figure 8.1 charts the frequency of anti-nuclear protest events after the Fukushima meltdown. The two largest events in terms of observed numbers of participants occurred in Tokyo in September 2011 and September 2012, with 60,000 and 75,000 protesters assembled, respectively. Interestingly, these citizen challenges did not, for the most part, arise from the established environmental groups in Japan. Anti-nuclear and international environmental groups played important organizing roles in these protests, but the vast majority of domestic environmental groups in Japan were silent. As mass protests continued over the course of a year and much of Japanese society grew critical of the hazards of the nuclear industry, a review of news reports, photos, and lists of event endorsers revealed a striking lack of environmental organizations (EOs).[3] Indeed, established environmental organizations played a very marginal, almost negligent role in these mass campaigns. As evidence

[1] As of 2011, Japan had fifty primary core reactors in operation and was projected to meet 40% of the country's electricity needs through nuclear power by 2017. Immediately following the meltdown, and amid growing protests across the country, the Japanese government and nuclear power industry temporarily closed down all electric power generation from nuclear sources. These temporary initiatives reflected a heightened legitimacy crisis, prompting a first-ever investigation into the industry. The commissioner charged with the investigation concluded that Fukushima's meltdown "was a profoundly man-made disaster," reflecting what the New York Times reported as "a preventable disaster rooted in government-industry collusion…" (Tabuchi, July 5, 2012). See Perrow [7] for an elaboration of the nuclear crisis.

[2] *USA Today*, reporting via the Associated Press from Tokyo, captured this common refrain [8].

[3] Our assertion stems from a comprehensive examination of all of the web-active groups in the Environmental Restoration and Conservation Agency [9] database, a national survey of all environmental organizations in Japan, including over 4800 groups. Among members of the anti-nuclear coalition that organized the national protests, only 4 organizations listed in the [9] database were found.

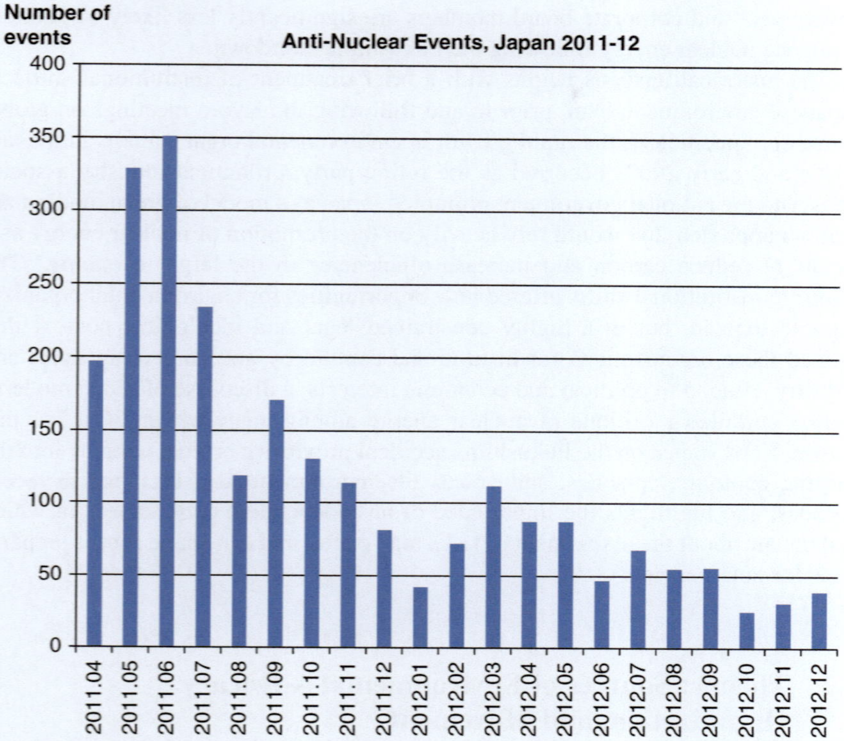

Fig. 8.1 Anti-nuclear protest events post-Fukushima. Note: Source data were obtained from "Anti-nuclear Power Movement Events Calendar," recoded, and then summed up by month. (http://datugeninfo.web.fc2.com/index.html). Reprinted with permission from Oxford University Press

of the nuclear contamination grew and over 100,000 Fukushima residents were evacuated from the Level 7 meltdown, what kept EOs from joining the national denouncement of nuclear energy amid the largest environmental crisis in the country's history?

By rethinking recent Japanese environmental movement history and incorporating survey data and board of director membership data into several multivariate analyses, this methodological combination identifies both historical and political-organizational factors that drove silence, and, by contrast, the relatively rare denouncement of nuclear energy by EOs after the meltdown. This approach builds on theory as well. Accounts of environmentalism and environmental policy in Japan stress the strength of bureaucrats in a paternalist state and the power of large corporations to constrain political opportunities for social movements generally, and environmentalism particularly [2, 11–15]. To assess this general theoretical argument, the combination of methods—which operationalize both organizational-level determinants of EO behavior and inter-organizational network factors—provide empirical conclusions from our statistical results: environmental groups with government and corporate board members are significantly less likely to publicly denounce nuclear energy following the Fukushima meltdown.

The historical analysis begins with a brief treatment of institutional shifts in Japanese environmentalism, prior to and following the Kyoto meetings on global warming. Specifically, the rapid growth in environmental organizations in the late 1990s and early 2000s occurred as the ruling party, prominent industry associations, and the national government promoted Japan as a model of green modernization, an approach that would rely heavily on the promotion of nuclear energy as a means to reduce carbon and increase efficiencies in the larger economy.[4] The resulting institutional shifts offered new opportunities for environmental organizations to expand, but in a highly constrained legal and ideological context that primed these organizations for institutional capture by government agencies and industry. Hinged to political and economic interests, a discourse of green modernization curtailed a critique of nuclear energy among many newer EOs. For this reason, EOs' stance on the Fukushima accident provides a critical window into the varying contours, episodes, and voices of environmentalism in Japan in recent decades, and highlights the importance of an independent civil society in which real debate about the assessment of risks may occur, and can shape robust preparations for emergencies.

8.2 Theory: Sources of Environmental Advocacy, Organization, and Movement

8.2.1 Environmental Identities

As complex organizations, EOs are not simply reacting to structural stimuli, but instead are constantly developing shared subjective interpretations of reality through a process of actively engaged sense-making about environmental problems, though within a definite historical context [16, 17]. The identity of an EO can be observed both in what is proclaimed in its mission statement, as well as what members of the organization actually do [18]. Identity and strategy are thus theoretically linked through this ideological component of EOs, as identity influences what issues are pursued, the channels and methods of resource acquisition [19], the tactics that the EO endorses [20], as well as the organizational structure that the EO employs.

While the identities and issue-focus of Japanese EOs vary considerably, a definite set of patterns are reported in the sociological literature. One important distinc-

[4]As Miyadai [13] concluded, the promotion of nuclear energy as carbon neutral is misleading: "If we look just at the process of generating electricity, it appears that there is no carbon dioxide emission. But seen in light of the entire process—including mining, refinement and enriching of uranium; transportation; building of power plants; decades of cool temperature storage; and more than 10 years of plant closure—such claims turn out to be pure exaggeration" (p. 99).

tion is made between a first wave of anti-pollution campaigns in the 1960s and 1970s from the later conservation initiatives that rose to prominence in the late 1970s and 1980s [12, 15, 21, 22]. Though distinct environmental identities emerged in both periods, the focus of Japanese environmentalism was overwhelmingly local [23]. Schreurs [15] points out that environmental advocacy for global issues was largely nonexistent prior to the 1980s, though a shift in environmental identities began as more groups formed with a focus on national and global ecological risks in the 1990s (see also [24, 25]). Where there was anti-nuclear activism, as Schreurs [15] argues, it too was more focused on local siting concerns, unlike the more national or global campaigns against nuclear energy in Germany or the USA (see also [26–29]).[5] We expect EOs with a focus on local environmental concerns, as well as conservation and park maintenance issues, to remain silent on the nuclear meltdown, while anti-pollution and internationally oriented groups denounce the disaster.

8.2.2 *Environmental Organizations and Resource Mobilization*

Prior to the 1990s, scholars agree that Japanese environmentalism did not develop large, national level movement organizations as occurred in other industrialized countries [12, 15, 30, 31]. Broadbent and Barrett [21] characterize EOs in Japan, and social movements in general as "very much on the periphery of institutional power, under-funded, under-supported, and disenfranchised" (p. 73). The organizational demography of the environmental movement consisted of numerous small organizations focused on town and regional, prefecture-based anti-pollution and conservation campaigns [24]. Changes in the organizational composition of the environmental movement in Japan since the early 1990s reflect improved resource mobilization capacities [24].

EOs in Japan are situated across a range of potential financial, institutional, and membership constraints. Examining the organizational dependencies of an EO, for example, is likely to be a critical determinant of its behavior [32]. Further, because EOs are unlikely to advocate positions opposed to their allies or compromise embedded ties, we consider these factors relevant [18]. Additionally, accepting financial assistance from state agencies tends to reduce the volatility of their advocacy, as those connections frequently leave EOs with a "loss of autonomy or independence, particularly [with] dilution of the sector's advocacy role" ([33]: 103; see also [12]). Contrarily, environmental organizations with fewer dependencies on state bureaucracies are likely less inhibited about their strategic choices [34].

[5] Local anti-pollution organizations maintained affinities and ideological resonance with the local opposition to the siting of nuclear power plants. The shared concern over the exposure of hazards at the local level is apparent in our results presented below. Among other results, we report that, net other factors, EOs further from the site of the Fukushima Daiichi nuclear power plant are less likely to denounce nuclear energy after the crisis.

Independent of structural dependencies, as an EO expands financially, it faces pressure to professionalize from three directions: (1) the increasing complexity of environmental legislation necessitates the expertise of a specialized paid staff, (2) rapidly expanding memberships require additional management, and (3) accountability becomes necessary to maintain their nonprofit status [35]. Pressures associated with increased finances and professionalization of EO leadership tend to constrain political action. In general, we expect few EOs with larger budgets to denounce nuclear energy post-Fukushima.

8.2.3 National and Transnational Political Opportunity Structures

Understanding environmental movement dynamics in Japan requires a specific conceptual framework for grasping the relationship between domestic political opportunities, state structures, and elite unity. The social movement literature is clear on this point: political opportunities for movement mobilization vary by the degree of political openness in the state, the relative unity of elites, and the availability of elite allies in the state or broader society [36, 37]. Broadbent's [12] scholarship provides such a framework for thinking about how the relatively strong state networks and a "ruling triad," or what Reed [38] refers to as the "triple control machine," interact to constrain political opportunities for environmental movement mobilization in Japan. Broadbent's [12] research explains how local environmental organizations in Japan, like other community organizations, are embedded in horizontal and vertical networks that inhibit autonomy and outward growth. In contrast to western NGOs, Broadbent argues that community organizations are bound by a social hegemony directed by clientelistic ties and the authority of senior leaders. These organizational patterns, Broadbent suggests [39], began to unravel as the dominant triad of party, state, and business faced a series of crises in the 1990s. New opportunities for community organization appeared in 1993 as the ruling party lost power and new grassroots campaigns grew [40]. A wave of new EO-formation occurred in this window of opportunity from the mid-1990s into the early 2000s. Though the sheer number of EOs increased, questions remain as to any differentiations among these EOs, especially with respect to their relative autonomy from the dynamics of social hegemony explained by Broadbent. Broadbent and others suggest, and we explore in detail elsewhere [41], that these newer EOs arose in a contradictory context, where many emerged with greater autonomy and others conformed to the longer pattern of government organized NGOs. *We expect evidence of a constricted political opportunity structure to be found in the networks between EOs and the political and industrial establishment*, which should correlate with the silence of EOs on nuclear risks after the Fukushima accident.

8.3 Grassroots Mobilizations and State Facilitated Environmentalism in Japan, 1950–2000

While business and state leaders had been actively encouraging the development of nuclear power plants and nuclear fuel reprocessing facilities for years, particularly after the second oil crisis in 1979, the Atomic Energy Commission of Japan (AEC) and the Ministry of Education, Culture, Sports, Science and Technology (MEXT) began to insist early in the 1990s that nuclear power was essential for mitigating climate change [42, 43]. Even as the anti-nuclear movement remained distinct from the larger environmental movement, focused largely on fights over local siting of new nuclear reactors, several national level anti-nuclear coalitions emerged (see [26]). This new politics of global warming would, however, change the terrain for both the anti-nuclear and environmental movement (Fig. 8.2).

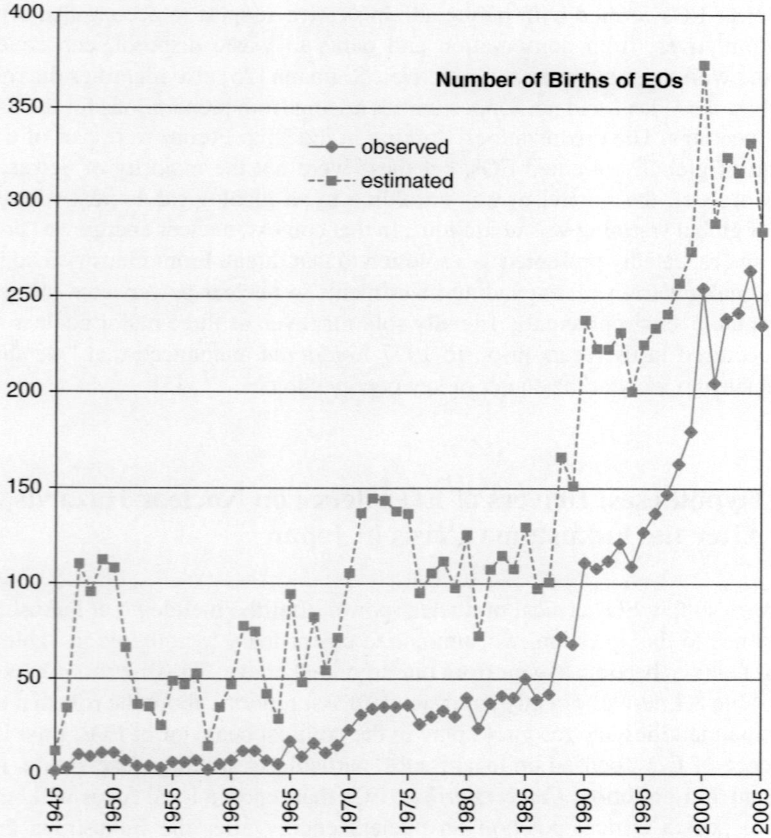

Fig. 8.2 Founding of environmental organizations in Japan, 1945–2005. Note: Reprinted with permission from Oxford University Press

New domestic political opportunities were opening for environmentalism at the same time that national elite sought domestic allies for asserting Japanese leadership in a high-tech vision for a carbon-efficient model of modernization. While Japanese EOs and other NGOs mobilized for the Third Conference of Parties (COP3) to the UN Framework Convention on Climate Change (FCCC), held on December 1–11, 1997 in Kyoto, nuclear power was at the center of the carbon reduction discussions. Hosting the COP3 meetings reinforced a view among national elite and local environmentalists that global warming was the problem and nuclear energy was the solution.

As legal and institutional changes expanded opportunities for new EOs to obtain funding for green projects with "local partnership" funds, avenues for government officials to "retire" on the boards of EOs grew as well. Combined with the "golden parachutes" for government officials [44], funding from public agencies, and the active program to open nonprofit status to more groups, many of the environmental organizations formed at this time arose as part of a *state facilitated environmentalism*. Newer EOs formed with public administrative support to accomplish official policy initiatives, from conservation and parks to waste disposal, consequently identified with those government initiatives. Reimann [25] also identifies the role of new allies, new ideas, and new opportunities arising from preparations for the Kyoto COP3 meetings. The organizations working in the Kiko Forum were part of a new segment of globally oriented EOs, but these were not the majority of newer EOs [25]. Moreover, their activism was unfolding in an ideological context where the threat of global warming was ubiquitous. In that context, nuclear energy was prominently and repeatedly presented as a solution to that threat. From industry leaders to the national government, expanding investments in nuclear power were offered as an acceptable, environmentally friendly solution, even as three major nuclear accidents occurred in the years prior. In 1997 *Keidanren* announced that "we should place nuclear energy at the center of key energy choices…" [45].

8.4 Hypotheses: Drivers of EO Silence on Nuclear Hazards After the Fukushima Crisis in Japan

Why were so few EOs critical of nuclear power after the meltdown at Fukushima? In response to this question, we summarize our primary hypotheses in Table 8.1. Several features become salient from our argument above. The first set of hypotheses in Table 8.1 derives directly from standard assumptions about the role that environmental identities and resources play in the political behavior of EOs. First is the prevalence of EOs focused on local issues, particularly around conservation, recycling, and anti-pollution. *Ceteris paribus*, EOs that report a local focus will be less likely to post a critical position on nuclear energy after the Fukushima crisis. Second, we also expect that EOs reporting as their primary concern "conservation" and "recycling" to be less likely to denounce the risks of nuclear contamination

Table 8.1 Hypothetical drivers of environmental organization denunciation of nuclear power after the Fukushima crisis in Japan

Identity, resource, and institutional hypotheses
• EOs that report a local (as opposed to national or global) focus in their mission will be less likely to denounce nuclear power post-Fukushima (−)
• EOs with identities that are single-issue focused on conservation (reforestation or beautification), recycling, or consumer protections will be less likely to denounce nuclear power post-Fukushima (−)
• EOs with identities that focus on anti-pollution and global environmental problems will be more likely to denounce nuclear power post-Fukushima (+)
• EOs with larger budgets will be less likely to denounce nuclear power post-Fukushima (−)
• EOs found in the decade after the Kyoto meetings will be less likely to denounce nuclear power post-Fukushima than those formed prior (−)
Network hypotheses
• EOs that have members of national or local government agencies on their boards will be less likely to denounce nuclear power post-Fukushima (−)
• EOs with corporate executives on their boards will be less likely to denounce nuclear power post-Fukushima (−)
• EOs with independent citizens, members of cooperatives, or other NPOs on their boards will be more likely to denounce nuclear power post-Fukushima (+)
• EOs with scholars or journalists on their boards will be more likely to denounce nuclear power post-Fukushima (+)

after Fukushima. In contrast, we expect anti-pollution organizations to be more likely to adopt a critical view of nuclear energy after Fukushima. This is because the anti-pollution identity resonates more closely with the local anti-nuclear waste and facility siting campaigns over the years, as well as the observed risks of nuclear contamination. These concerns have been reported to resonate among the Japanese public in other periods, for example, following the Three Mile Island and Chernobyl accidents.[6] EOs identified with any number of global issues will tend to associate with ideas prevalent in the international environmental community, including concerns with nuclear risks, and will therefore be more likely to express a critical view of nuclear power after the Fukushima crisis. Concerning resources, we expect EOs with larger budgets to be less likely to challenge prevailing programs of the state and therefore remain silent on the nuclear crisis after Fukushima.

Combined with passage of the NPO law (1998) and state support for conservation and carbon reductions during and after the Kyoto meetings on climate change, we expect EOs found after 1998 to be less likely to denounce the hazards of nuclear energy after Fukushima. Environmental advocacy that is subsumed under official

[6] In August 1986, after the Chernobyl incidents (April 26), a survey of public opinion carried out by Asahi Shimbun reported that the percentage of opposing nuclear power plant development (41%) exceeded that of pro-nuclear response (34%). This was the first instance where opinions swayed against nuclear power since the survey started in 1978. In 1988, 20,000 people were rallied to an anti-nuclear gathering at Hibiya Park in Tokyo [44]. For a comparison of environmental impacts of the Chernobyl and Fukushima disasters, see Steinhauser et al. [45].

policy goals, we reason, is less likely to oppose state energy and industrial policies. This is not merely a consequence of the founding date of the EO or the specific local partnership objective, but also a function of the political and organizational embeddedness of the EO governing board across the state, economy, and civil society. Specifically, we expect EOs with positive indicators of institutional capture—arising from the selective use of "golden parachutes" that place government or business members on the board of directors for the EO—to remain silent. A greater number of government officials and business leaders on the board of directors of an EO is likely to present political-organizational constraints on the identity and behavior of the EO. We therefore posit that EOs with board members from government agencies or corporations to be less likely to denounce nuclear energy after the Fukushima crisis. Conversely, EOs with positive indicators of civil autonomy—possessing citizens, journalists, or cooperative members on their boards—will be more likely to denounce nuclear energy after Fukushima.

8.5 Data and Analyses

Survey data were obtained from Japan's Environmental Restoration and Conservation Agency census survey of environmental nonprofits [10]. This survey collected a wide range of data on 4855 EOs throughout Japan, including basic EO demographic data, numerous response items on the mission and principal activities of the EO, a categorical measure of annual revenues, headquarters location, and geographic scale of the EO's focus. ERCA survey data is collected only on registered or incorporated nonprofit organizations. An unknown number of environmental groups do not register or officially incorporate in Japan, meaning that our analyses do not apply to those groups. All EOs in the ERCA population of registered environmental nonprofit organizations were then sought out on the internet. We identified 2223 EOs with active, recently updated websites. Because of missing data on several key survey response items, 38 of these EOs were excluded from the analyses, leaving 2185.

We also sought to collect additional, publicly accessible financial and board membership data on each EO. Unfortunately, access to this data proved limited and difficult to obtain. Our efforts nonetheless resulted in a substantial, non-random subsample of 278 EOs from the 2185 web-active EOs. Combined with the ERCA survey data, this smaller subsample includes detailed information on the board members of the EOs, enabling the operationalization of board interlocks with numerous institutions in Japan. Based on board member biographic and affiliation data, we then coded network indicators between these EOs and government agencies (national and prefectural), media, other nonprofits, businesses, and universities. Together, the research design operationalized several theoretically salient factors driving environmental advocacy and policymaking in Japan. Our multivariate models and network indicators combined to make better sense of the responses by environmental organizations to the aftermath of Fukushima (for a detailed discussion of the data and operationalization of variables see [41]).

Below, we conduct two logistic analyses on the likelihood of EOs denouncing nuclear energy after Fukushima, first testing conventional variables of EO behavior in the larger sample ($n = 2185$) followed by a second regression with the smaller sample ($n = 278$) that incorporates the organizational network variables.[7] Our results highlight the impact of the political-organizational embeddedness of EOs with the state, business, and civil society and point to the historic role played by the political and industrial establishment in shaping Japanese environmentalism prior to and during the Kyoto COP3 meetings. Overall, the analyses reinforce an interpretation that considers how historical conditions shaped environmental advocacy in Japan in a way that privileged nuclear energy in the face of a narrative on the threats from global warming and climate change.

Our final analysis consists of a graph-theoretic visualization of the network data. In addition to the EO board member data, which is incorporated into the second logistic analysis, we construct a separate matrix of official affiliations between EOs, major government ministries, numerous universities, and private corporations. A graph visualization method is used to construct an image of these organizations and their ties in a manner that spatially distributes similar pairs of organizations proximate to one another. This graph representation provides a visual heuristic of the underlying network structure (Table 8.2).

8.6 Results

Model 1 in Table 8.3 presents the results of our first logistic regression on the odds of environmental organizations (EOs) publicly denouncing nuclear energy after the Fukushima crisis. We present both the log odds estimate and the odds ratio in Table 8.3. This first model examines organizations who completed the survey data from ERCA and had websites for coding their public positions on the Fukushima meltdown ($n = 2185$). Three broad patterns are immediately apparent. First, the environmental identities reported by the organizations are associated with different outcomes toward nuclear energy, post-Fukushima. Model 1 offers important validation of previous scholarship explaining the important distinctions among the

[7] Both of our samples differ with respect to financial size and locality from the population of all EOs in the ERCA national survey. With regard to our larger sample ($n = 2185$), only EOs with active, recently updated websites were included. Second, with our smaller sample ($n = 278$), because the ERCA data provides only limited information on the sources of organizational finances and no board membership details for EOs, this sub-sample was restricted to organizations where such revenue and board membership data was accessible. Each of these approaches resulted in samples that are comprised of larger environmental organizations, on average, and organizations that are slightly more inclined to work on national level issues as opposed to primarily local or neighborhood concerns. The differences on size and locality are statistically significant and warrant caution in the generalizability of our results below.

Table 8.2 Results of coding EO websites following the Fukushima meltdown, October 2012–April 2013 ($n = 2223$)

Coding for EO responses to the Fukushima meltdown	0: Express nothing about nuclear energy even in reference to the tsunami and humanitarian crisis	1: Show the facts/no oppositional stance	2: Hold conference or host study groups/ publish in bulletin or blog	3: Insist to revise energy policy/place responsibility on society in general	4: Oppose but insist gradual phase out/place responsibility on nuclear industry or government	5: Strongly oppose, insist to stop plants immediately/condemn TEPCO and/or government
Number of EOs	2075	35	41	22	26	24
%	93.3	1.6	1.8	1.0	1.2	1.1
Mention earthquake or tsunami disaster			Mention nuclear power plant crisis			
	Count	Percent		Count	Percent	
Mentioned	391	17.6	Mentioned	148	6.7	
Did not mention	1832	82.4	Did not mention	2075	93.3	

Note: Reprinted with permission from Oxford University Press

Table 8.3 Odds of denouncing nuclear energy post-Fukushima

Predictor variable	Model 1			Model 2		
	Estimate	Pr > ChiSq	Odds ratio	Estimate	Pr > ChiSq	Odds ratio
Analysis of maximum likelihood estimates						
Intercept	−3.3897	<0.0001		0.2066	0.7224	
Anti-pollution and global environmental identity	1.1471	<0.0001	3.149	1.1259	0.0003	3.083
Conservation and beautification identity	−1.1886	<0.0001	0.305	−1.1814	0.0036	0.307
Consumer protection identity	0.7260	0.0049	2.067	1.2328	0.0046	3.431
Locality (1, village/town; 7, international)	0.1680	0.0025	1.191	0.1282	0.1410	1.137
Annual budget (categorical 0-4)	0.0471	0.6159	1.048	−0.6091	<0.0001	0.544
Formed before 1999=0, else 1	−0.9980	<0.0001	0.369	−0.6984	0.0488	0.497
Max-rescaled R-square	$n = 2185$	0.1515		$n = 278$	0.2808	
Ass. pred. prob. and obsd. resp.		77.7			78.5	

Note: Reprinted with permission from Oxford University Press environmental identities among EOs in Japan [15]. This scholarship established our expectations that EOs with a more global outlook on environmental problems would adopt a more oppositional stance to the risks of nuclear energy. Environmental organizations that primarily identify with anti-pollution and global environmental concerns are 215% more likely to publicly denounce nuclear energy. EOs with a consumer protection identity are also about 107% more likely to denounce nuclear energy post-Fukushima. This is in contrast to those EOs reporting a primary focus on local beautification and conservation issues, who are 3.28 times less likely to denounce nuclear energy.

Second, the founding era of the organizations significantly impacts whether or not they would make a public statement denouncing nuclear energy post-Fukushima. Environmental groups formed in 1999 or after—following the 1998 NPO Law and the concerted campaign to promote nuclear energy as a solution to global warming—are *significantly less likely* to oppose nuclear energy. The inverse odds ratio (1/.369) indicates that EOs formed after 1998 were 2.71 times less likely to denounce nuclear energy after the Fukushima crisis. These results invite us to explore how silence on nuclear risks is nested in overlapping fields of state, corporate, and environmental politics, not just stated environmental identities or priorities.

Third, the geographic focus of the EO—found in our "locality" variable—positively increases the likelihood of denouncing nuclear energy by about 19% for each increase in the scale of geographic focus. EOs with wider geographic concerns are more likely to oppose nuclear energy after the crisis. Finally, contrary to our expectations, the budget size of the EO failed to approach statistical significance.

For reasons found in the literature, it was desirable to test additional factors driving opposition to nuclear energy. Because neither the ERCA data nor most EO websites offered detailed financial or board member information about EOs, we limit the next to only those ERCA surveyed organizations that provided public information about their board membership. This greatly diminished the sample size. The results of the regression on this smaller sample are presented alongside Model 1 in Table 8.3 under Model 2. The results remain roughly analogous to those found in Model 1, though the size of the EO's annual budget becomes significant. In Model 2, for each increase in budget size, an EO is 1.84 times less likely to denounce to nuclear energy. Consistent with our reasoning from the literature on environmental movements, larger organizations, net other factors, will tend to moderate their political positions. The locality variable dropped out of significance in this model and each of the three identity variables retained significance in the direction found in Model 1, though consumer identity saw a significant increase in impact. The contrast between Model 1 and Model 2 confirm that our two samples are modestly different with respect to the size and geographic scope of EO operations.

Our next analysis is presented in Table 8.4. We introduce the count variables of the types of board connections to outside organizations among the sample of EOs. For the ease of interpretation, the odds (and inverse odds) ratios are reported in Table 8.4. This smaller sample offers results analogous to Model 2 in Table 8.3,

Table 8.4 Odds of denouncing nuclear energy post-Fukushima, with political-organizational variables

Predictor variable	Odds ratios	Inverse odds ratios	Standard errors
Scholar/journalist on board	1.320*		0.0761
National or local govt. on board	0.653*	1.531	0.1361
Corporate exec. on board	0.770*	1.299	0.0668
Coop, NPO, or citizen on board	1.110*		0.0388
Anti-pollution and global env. identity	3.784*		0.3914
Conservation and beautification identity	0.264*	3.788	0.4788
Annual budget (categorical, 0–4)	0.589*	1.698	0.1852
Formed before 1999=0, else 1	0.224*	4.464	0.4288

Significance tests for odds ratio estimates derived from 95% Wald confidence intervals, $*P < 0.05$; R-square, 0.4086; Max-rescaled R-square, 0.5756; *N*, 278
Note: Reprinted with permission from Oxford University Press

though the consumer protection identity and the locality variables are not included because they drop out of significance with the introduction of the board variables. The four variables that remain significant from Model 2 also operate in the same direction after adding the four board count variables. EOs identifying with anti-pollution and global environmental concerns are 278% more likely to denounce nuclear energy and, in contrast, EOs identifying with conservation and parks beautification are 3.8 times less likely to denounce nuclear energy. For each increase in budget size, we find a 70% decrease in chance the EO will denounce nuclear energy.[8]

Results in Table 8.4 affirm the significance of our theoretical propositions concerning the institutional dependency and organizational embeddedness of EOs. The presence of a "golden parachuted" official on a local EO board is negatively associated with a critical position on nuclear energy. Indeed, for each additional government or corporate official, respectively, on their board EOs are 1.5 or 1.3 times less likely to denounce nuclear energy. Alternatively, those EOs with board members who are identified as scholars from universities, journalists, members of cooperatives, local citizenry or other nonprofit organizations (NPOs) are more likely to denounce nuclear energy. Each additional scholar or journalist increases the likelihood of denouncing nuclear power by 32% and each additional member of a cooperative, another nonprofit or unaffiliated citizen, increases the likelihood of denouncing nuclear energy by 11%. Finally, our historical founding variable again confirms the significant cohort effect on EOs formed after the 1998 NPO law. In this sample, EOs formed after 1998 were nearly 4.5 times *less likely* to denounce nuclear energy.

[8]A bivariate correlation between budget size and the number of outside board members (centrality) was 0.23, showing that financially larger EOs in this sample are more likely to have larger boards and therefore more outside directors. Net other factors, the larger EOs have more ties to government and business, though many large EOs maintain ties to other civil society institutions (unions, universities, coops, etc.).

8.6.1 Administrative and Corporate Embeddedness of Environmental Advocacy

Why do the network connections to government agencies and private corporations so significantly reduce the odds that an EO will denounce nuclear energy after the Fukushima crisis? Building on Broadbent's [12] concept of a "ruling triad," we present a network heuristic that maps relationships between a large subsample of EOs, corporations, and government agencies.

Beginning in the early 1990s, state leaders actively facilitated a particular variant of environmental advocacy in Japan. These initiatives coincided with an emergent elite consensus that Japan's national and international stature could be boosted by the promotion of green and efficient modernization. This vision relied on the promotion of a nuclear safety myth [3] as a negation of the environmental costs of the nuclear industry, which help power this relatively low-carbon economy. Existing state structures, revised laws, and other initiatives by business and government facilitated a rapid growth of environmental organizations at this time, often funded directly via local partnership initiatives aimed at conservation, parks, and waste management. This state facilitated environmentalism advanced a *proscriptive frame* that nuclear power could safely advance national economic development while protecting the country (and world) from the threat of global warming and climate change. In this way, state facilitated environmentalism constrained EOs from reflexively weighing the hazards and risks of nuclear energy, even after the Fukushima crisis. Our regression model in Table 8.4 validates this narrative.

To further examine the political-organizational embeddedness of EOs, we present data on these overlapping sectors in Fig. 8.3. Using coordinates from the Fruchterman-Reingold algorithm in NodeXL, Fig. 8.3 plots the affiliations among a subsample of EOs, government agencies, universities, foundations, and corporations that result from common personnel on the boards of the respective entities. In this figure, each node depicts an organization or agency, and an arc between two nodes indicates at least one common member shared between the two.

Figure 8.3 is a snapshot of the political-organizational embeddedness of our subsample of EOs. In reality, these networks are far more expansive. It is apparent that many EOs accept board members, including full-time and paid ones from METI, MOE, MLIT,[9] MHLW,[10] and other government agencies. This figure also reveals dense connections between EOs and various industries, including the nuclear energy industry, by means of board appointments. Five major electric utility corporations, which own the nuclear reactors throughout Japan, have direct ties to EOs in our subsample. Major EOs are immersed in this network, including prominent ones such as the "Institute for Global Environmental Strategies," "Japan Environmental Association," "Japan Environmental Management Association for Industry," "Global Environmental Forum," "Japan Environmental Education Forum," and

[9] Ministry of Land, Infrastructure, and Transport.

[10] Ministry of Health, Labor, and Welfare.

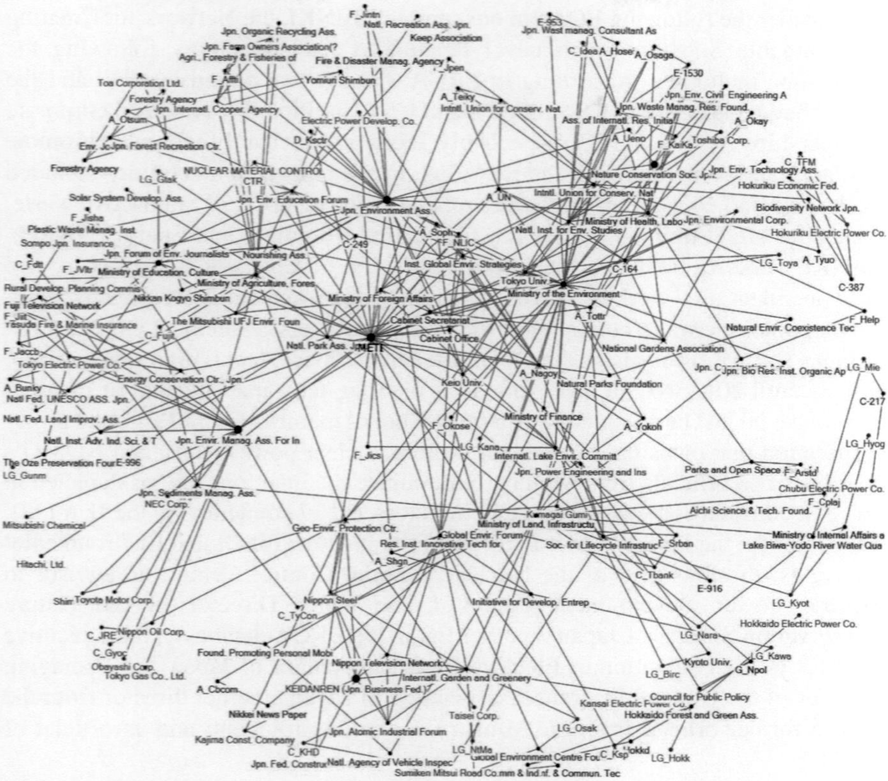

Fig. 8.3 The political-organizational embeddedness of EO Subsample, 2012. Note: A_: University, College; B_: Bank and Insurance; C_: Corporation; D_: Independent administrative agency; F_: Foundation; G_: Government; LG_: Prefectural (Local) Government; M_: Media; E-, C-, K-, H-, T-, S-, N-: Environmental organizations (the first letter indicates the location of the EO). Reprinted with permission from Oxford University Press

"Research Institute of Innovative Technology for the Earth." These organizations have multiple board connections with government agencies, corporations, and other EOs. The density of network overlap between industries, government agencies, and environmental organizations is thus interpreted as a highly constricted political opportunity structure, one not likely to support a critique of the nuclear energy industry. EOs embedded in this political-organizational network inhabit a context that inhibits a reflexive critique of the risks of nuclear power in Japan.

As a result, leading EOs were a minor part of the citizen protests against nuclear energy after the Fukushima crisis. We conclude that national energy policy priorities are not determined by ecologically reflexive responses to disaster (human or environmental) but are instead propelled by an interconnected complex of industry, government, and a state facilitated nonprofit sector. The distortion of ecological discourse is the result (see [16]).

Consider the following EO from our sample, GENKI, the Network for Creating a Sustainable Society, which never denounced nuclear energy following the Fukushima meltdown. According to ERCA, the EO was found in 1996 and the annual budget is between US $100,000 and US $1 million. Personal and corporate memberships are 400 and 20, respectively. It is a pro-nuclear EO found to promote and facilitate waste reduction and recycling. In one report, their activities included workshops and educational initiatives related to the disposal of radioactive waste, funded by METI through the Agency for Natural Resources and Energy (ANRE). GENKI works to convince local communities to accept radioactive waste, and their president also worked for government agencies and similarly oriented EOs. For example, she was a councilor of the Nuclear Waste Management Organization of Japan (NUMO) and also a director of the Japan Environment Association (JEA), at least until 2012. NUMO was found for the long-term management of radioactive waste. NUMO's president was a former board member for TEPCO (the utility corporation that owns the Fukushima Daiichi nuclear power plant) and NUMO's board has two officials from METI. To illuminate the interconnections depicted in Fig. 8.3, consider some of the other 21 directors and 17 councilors in the JEA [48]. They include: the Former Chief Director of the Institute for Global Environmental Strategies; the President of the Nuclear Material Control Center; an advisor to Electric Power Development Co., Ltd; the Chief Director of the Nature Conservation Society of Japan (another EO in the ERCA database); the Executive Director of the Federation of Electric Power Companies of Japan; the Managing Director of the General Insurance Association of Japan; another director from the JEA; a former official from the Ministry of the Environment; and an official of METI.

In contrast, consider an EO that denounced nuclear energy after the Fukushima accident. The Center for International Cooperation (JANIC) was primarily engaged in international training of NGO members engaged in environmental protection initiatives. After the meltdown of Fukushima, they devoted much of their efforts to supporting tsunami-stricken areas, and on October 4, 2011, the Chairperson of JANIC, citing environmental and human hazards, Masaaki Ohashi declared, "there is no room left to accept nuclear power generation in Japanese society" [49]. Except for one board member from the Japan International Cooperation Agency (JICA), which is an agency of the Ministry of Foreign Affairs, their board contains no other officials from government agencies. Their board also contains one professor, a member of the Japanese Organization for International Cooperation in Family Planning, and the Director of International Affairs from the Japanese Trade Union Confederation. Their revenue sources are also revealing, with 20% obtained from multi-denominational religious foundations, labor organizations, and other organizations as grants, 45% from a variety of religious organizations and companies as contributions, and 24% from the Ministry of Foreign Affairs and its agency's (JICA) commissioned project. JANIC was found in 1987.

8.7 Conclusion

Despite the threat of future nuclear accidents from earthquakes, environmental organizations in Japan have largely remained silent on the risks of nuclear energy following the Fukushima disaster [8, 50]. This was found to be at least partly explained by meaningful variations in their identities, resources, and their embeddedness in political-organizational networks. Accepting nuclear energy production as a mitigating strategy against global warming does not necessarily preclude a critical assessment of the risks associated with nuclear power, and indeed will be an essential caution as alternatives to carbon-intensive energy sources are devised. The adoption of this attitude among many EOs meant a tacit approval for restarting nuclear plants and continued to position Japan as the rhetorical leader in the global struggle to reduce greenhouse emissions. Even when articulated by major environmental organizations (e.g., the Hokkaido Environmental Foundation (2011), who stressed that the expansion of nuclear activities must be constrained and existing plants decommissioned) as an imperfect solution that must eventually be replaced, few EOs insist on a permanent and complete decommissioning of nuclear reactors.[11]

The hesitance of many EOs to denounce the nuclear accident in Japan was partially rooted in a constrained political opportunity structure for anti-nuclear claims arising since an historical conjuncture beginning in the mid-1990s. Many EOs were formed at the height of a national campaign against global warming and an ideological context that minimized the risks of nuclear energy relative to the very real threat of global warming. With greater legal access to nonprofit status and support from major government funding, their fields of action were shaped by political-organizational networks to favor a status quo energy and environmental policy. Despite being hailed as a source of civil society opening in Japan, the 1998 NPO Law also set new normative and institutional constraints on EOs formed in the years following its passage. Thus, to understand EO silence *and* voice in the wake of the Fukushima disaster, we must look beyond standard organizational profiles and consider the historical and network context where new EO cohorts arise. Such an approach can inform risk preparedness by properly understanding the conditions that help or hinder dialogue among state, economic, and civil society actors.

References

1. Hasegawa K. Memorial address to commemorate the first anniversary of the great east Japan earthquake. 2012. http://www.zenkoku-net.org/topics/topic120309.php. Accessed 29 Apr 2014.

[11] It is in this context that the president of the Japan Network for Climate Change Actions, Koichi Hasegawa, framed the post-Fukushima era as a turning point of Japanese energy policy [1].

2. Hasegawa K. Facing nuclear risks: lessons from the Fukushima nuclear disaster. Int J Jpn Sociol. 2012;21:84–91.
3. Funabashi H. Why the Fukushima nuclear disaster is a man-made calamity. Int J Jpn Sociol. 2012;21:66–75.
4. Kingston J. Abe's nuclear renaissance. Crit Asian Stud. 2014;46(3):461–84.
5. Beck U. Risk society: towards a new modernity. London: Sage Publications; 1992.
6. Noggerath J, Geller RJ, Gusiakov VK. Fukushima: the myth of safety, the reality of geoscience. Bull Atom Sci. 2011;67(5):37–46.
7. Tabuchi H. Inquiry declares Fukushima crisis a man-made disaster. 2012. http://www.nytimes.com/2012/07/06/world/asia/fukushima-nuclear-crisis-a-man-made-disaster-report-says.html?_r=3&. Accessed 19 Feb 2014.
8. Perrow C. Nuclear denial: from Hiroshima to Fukushima. Bull At Sci. 2013;69(5):56–67.
9. Associated Press. Thousands march against nuclear power in Tokyo. 2014. http://usatoday30.usatoday.com/news/world/story/2011-09-19/japan-anti-nuclear-protest/50461872/1. Accessed 19 Feb 2014.
10. ERCA (Environmental Restoration and Conservation Agency). 2007. http://www.erca.go.jp/jfge/english/. Accessed 29 Apr 2014.
11. Broadbent J. Environmental politics in Japan: an integrated structural analysis. Socio Forum. 1989;4(2):179–202.
12. Broadbent J. Environmental politics in Japan: networks of power and protest. Cambridge: Cambridge University Press; 1998.
13. Hasegawa K. A comparative study of social movements for a post-nuclear energy era in Japan and the USA. In: Broadbent J, Brockman V, editors. East Asian social movements; power, protest and change in a dynamic region. New York: Springer Science and Business Media; 2011. p. 63–79.
14. Miyadai S. Pitfalls of the 'nuclear power reduction' movement. Int J Jpn Sociol. 2012;21:98–107.
15. Schreurs M. Environmental politics in Japan, Germany, and the United States. Cambridge: Cambridge University Press; 2002.
16. Brulle RJ. Agency, democracy, and nature: the US environmental movement from a critical theory perspective. Cambridge: MIT Press; 2000.
17. Klandermans B. The peace movement and social movement theory. Int Soc Mov Res. 1991;3:1–39.
18. Dreiling M, Wolf B. Environmental movement organizations and political strategy. Organ Environ. 2001;14:34–54.
19. Knoke D. Organizing for collective action: the political economies of associations. New York: A. De Gruyter; 1990.
20. Benford RD, Hunt SA. Dramaturgy and social movements: the social construction and communication of power. Sociol Inq. 1992;62:36–55.
21. Broadbent J, Barrett B. The transformation of social movements and civil society. In: Barrett B, Fisher D, editors. Ecological modernization and Japan. New York: Routledge Press; 2005. p. 67–88.
22. Nishikido M. The dynamics of protest about environmental issues in Japan after WWII: using protest event data. J Environ Sociol. 2003;9:107–23.
23. Funabashi H. Environmental problems in post-war Japanese society. Int J Jpn Sociol. 1992;1:3–18.
24. Johnson EW, Saito Y, Nishikido M. The organizational demography of Japanese environmentalism. Sociol Inq. 2009;79(4):481–504.
25. Reimann KD. Building networks from the outside in: international movements, Japanese NGOs, and the Kyoto climate change conference. Mobilization Int Q. 2001;6(1):69–82.
26. Aldrich DP. Site fights: divisive facilities and civil society in Japan and the West. Ithaca: Cornell University Press; 2010.
27. Hasegawa K. The radioactive and industrial waste problem from a sociological perspective. J Environ Sociol. 2000;6:169–238.
28. Hasegawa K. Constructing civil society in Japan: voices of environmental movements. Melbourne: Trans Pacific Press; 2004.

29. Lesbirel H. NIMBY politics in Japan: energy siting and the management of environmental conflict. Ithaca: Cornell University Press; 1998.
30. McKean MA. Environmental protest and citizen politics in Japan. Berkeley: University of California Press; 1981.
31. Sasaki-Uemura WM. Organizing the spontaneous: citizen protest in postwar Japan. Honolulu: University of Hawaii Press; 2001.
32. Dreiling M, Lougee N, Nakamura T, Jonna RJ. Environmental organizations and communication praxis: a study of communication strategies among a national sample of environmental organizations. Organ Environ. 2008;21(4):420–45.
33. Salamon LM. Partners in public service: government/nonprofit relations in the modern welfare state. Baltimore: Johns Hopkins University Press; 1995.
34. Andrews KT, Edwards B. Advocacy organizations in the U.S. political process. Annu Rev Sociol. 2004;30:479–506.
35. Mitchell RC, Mertig AG, Dunlap RE. Twenty years of environmental mobilization: trends among national environmental organizations. In: Dunlap RE, Mertig AG, editors. American environmentalism: the U.S. environmental movement, 1970-1990. Washington: Taylor & Francis; 1992. p. 219–34.
36. Jenkins C. The politics of insurgency. New York: Columbia University Press; 1985.
37. McAdam D. Political process and the development of Black insurgency, 1930-1970. Chicago: University of Chicago Press; 1999.
38. Reed SR. Japanese prefectures and policy-making. Pittsburgh: University of Pittsburgh Press; 1986.
39. Broadbent J. Movement in context: thick networks and Japanese environmental protest. In: Diani M, McAdam D, editors. Social movements and networks: relational approaches to collective action. New York: Oxford University Press; 2003. p. 206–29.
40. Taguchi M. Gomi mondai hyakka II: Soten to tenbo. Tokyo: Shin Nihon Shuppansha; 1998.
41. Dreiling M, Lougee N, Nakamura T. After the meltdown: explaining the silence of Japanese environmental organizations on the Fukushima nuclear crisis. Soc Probl. 2017;64:86–105.. https://doi.org/10.1093/socpro/spw034
42. AEC (Atomic Energy Commission). White paper on nuclear energy. 1990. http://www.aec.go.jp/jicst/NC/about/hakusho/wp1990/sb1010101.htm. Accessed 29 Apr 2014.
43. MEXT (Ministry of Education, Culture, Sports, Science and Technology). 1994. http://www.mext.go.jp/b_menu/hakusho/html/hpaa199401/hpaa199401_2_073.html. Accessed 29 Apr 2014.
44. Terada R. The institutionalization of environmental NPOs and its impact on environmental movements. J Environ Sociol. 1998;4:7–23.
45. Japan Business Federation. A proposal for the way nuclear policy should be, July 22nd 1997. 1997. http://www.keidanren.or.jp/japanese/policy/pol139.html. Accessed 24 Jan 2014.
46. Asahi Shimbun Reporting Team. Nuclear power plant and media 2. Japan: Asahi Shimbun Publications, Inc; 2013.
47. Steinhauser G, Brandl A, Johnson TE. Comparison of the Chernobyl and Fukushima nuclear accidents: a review of the environmental impacts. Sci Total Environ. 2014;470(471):800–17.
48. JEA (Japan Environmental Association). Officers and Councilors. 2010. http://www.jeas.or.jp/english/director.html. Accessed 29 Apr 2014.
49. Ohashi M. JANIC's position statement regarding nuclear power plants and sustainable societies. 2011. http://www.janic.org/en/earthquake/news/information/janics_position_statement_regarding_nuclear_power_plants_and_sustainable_societies.php. Accessed 23 Dec 2013.
50. Perrow C. The next catastrophe: reducing our vulnerabilities to natural, industrial, and terrorist disasters. Princeton: Princeton University Press; 2007.

Chapter 9
Public Relations in Times of Nuclear Emergencies: Examples from a Medium-Sized Public University and a Small Austrian Municipality

Bettina Neunteufl

Abstract Surrounded by nuclear power plants, Austria forms a nuclear power-free enclave in the heart of Europe. In 1978, the Austrian population voted against the commissioning of the already built NPP Zwentendorf. Forty years later and after the shutdown of the ASTRA research reactor in Seibersdorf, the TRIGA Mark II reactor commissioned in 1962 at the Institute of Atomic and Subatomic Physics (ATI) of the Vienna University of Technology (TU Wien) remains the only nuclear reactor facility in Austria. Located in the Viennese Prater (Vienna's largest recreational park area), the said reactor has defined a significant fraction of the research and teaching activities of the ATI, including training for external partners. Together with the associated teaching and research infrastructure, it puts the TU Wien in a position to work with radioactive materials and ionizing radiation and to teach the safe and responsible handling of nuclear materials. An important complementary activity of the ATI is to conduct training workshops for international experts from the International Atomic Energy Agency (IAEA). This chapter aims at providing an insight into public relations and science communication strategies in the event of a nuclear emergency with special focus on both a personal and a university's perspective. Three case studies are discussed, in particular information management after the Fukushima crisis by the ATI, communication strategies during a delicate refueling procedure of a nuclear reactor, and perception of nuclear crisis management and communication by the mayor of a small Austrian municipality. When confronted by a massive nuclear crisis, a university can develop great strength in crisis communication and become a source of trustworthy and unbiased information to the public and the media.

Keywords Austria · Nuclear emergencies · Crisis communication · Public relations · Public university

---

B. Neunteufl (✉)
TU Wien, Vienna University of Technology, Service Unit of Public Affairs and Press Spokesperson, Vienna, Austria
e-mail: bettina.neunteufl@tuwien.ac.at

G. Steinhauser et al. (eds.), *Nuclear Emergencies*,
Current Topics in Environmental Health and Preventive Medicine,
https://doi.org/10.1007/978-981-13-8327-4_9

9.1 Introduction

An unexpected nuclear emergency can disrupt established communication structures and easily overstrain years of experience in public relations (PR) for stakeholders that are involved or suddenly become involved in a nuclear crisis. Not always, to say the least, is communication and PR in times of a nuclear crisis only about facts that are relevant to the (public) understanding of the crisis. In this chapter, I will outline some of the challenges that the PR office of Vienna University of Technology (TU Wien)—Austria's sole operator of a nuclear research reactor—had been facing in the course of the Fukushima crisis—and beyond. Together with some anecdotal personal experiences, I will try to illustrate the challenges that evolved in nuclear emergency communication from the perspective of TU Wien, Austria's largest university for higher education in technical sciences (approx. 30,000 students).

In the late 1970s, I grew up together with four siblings in a small village called Langau, which is located in northern Lower Austria very close to the border to former Czechoslovakia. In 1978, I learned to ride a bike at the age of three. It was just this year that Austrian citizens were called upon for the first time in the history of the Second Republic to vote in a referendum on the question whether or not the already completed nuclear power plant (NPP) Zwentendorf should go into operation. On 5 November 1978, 50.47% of Austria's population voted against the commissioning of the NPP Zwentendorf. Since then, this result has influenced the governmental policy in nuclear power. In December 1978, the so-called Nuclear Cuts Act [1] passed the parliament. The government mothballed the already completed NPP Zwentendorf which became a multi-billion dollar industrial memorial site. In addition, the afore-cited Nuclear Cuts Act put the nail in the coffin of the second nuclear power station planned in Stein/St. Pantaleon, Upper Austria. Since then, nuclear-generated electricity has been considered undesirable in Austria—but it is still being imported. Although there had been repeated attempts to reconsider the "No" to nuclear power, at the latest with the nuclear disaster in Chernobyl 1986 the topic became undebatable in Austria. In 1986, I was eleven years old, had a bigger bike, became an aunt for the first time, and was still living next to the strictly closed border to Czechoslovakia. What was happening on the other side of the "iron curtain," the children on our side of the border did not know at all.

Austria has also been a member of EURATOM (European Atomic Energy Community) since joining the European Union (EU) on 1 January 1995. Already before and on Austria's accession to EURATOM, there were strong ressentiments among the Austrians population about the membership in EURATOM and concerns about nuclear power in general. In February 2018, Austria sued the European Commission (EC) at the European Court of Justice in Luxembourg for allowing Hungary to expand its NPP at Paks. The NPP at Paks on the Danube about 100 km south of Budapest has been in operation since 1982. The Austrian government asked the court to annul the EC's resolution approving the expansion. Similarly, Austria had acted against the construction of the British NPP Hinkley Point C in 2015.

The obviously critical Austrian attitude against nuclear power is reflected in studies by the European Commission. The Directorate General for Energy and Transport of the EC launched Special Eurobarometer reports on "Attitudes towards radioactive waste" and public opinion of "Europeans and Nuclear Safety" following three former studies on radioactive waste carried out in 1998, 2001, and 2005. According to the Special Eurobarometer 297 [2], only 14 percent of Austrians are in favor of using nuclear power, "The lowest support for nuclear energy is, however, clearly found in countries that have no nuclear power plants. The least support for this type of energy is found in Austria, Cyprus and Greece, with around eight in ten respondents confirming that they are opposed to this type of energy." However, even Austrians are aware of the fact that global abandoning nuclear power would not be a realistic expectation. According to the Special Eurobarometer 324 [3], "Europeans have a moderate level of knowledge of nuclear issues: though few respondents knew that the European Union has the largest number of nuclear power plants in the world, they were more aware that nuclear waste is not exclusively produced by nuclear power plants. Similarly, Europeans continue to be unfamiliar with safety issues related to nuclear power plants. Only a quarter of citizens feel 'very well' or 'fairly well' informed, compared with three in four who feel 'not very well', or 'not at all' informed about the safety of nuclear power plants. This situation is almost identical to the one depicted in the previous survey." Asked about the role of media in the context of information the results show that "Europeans are critical of the information offered in the media about energy in general and nuclear energy in particular: almost two thirds of the interviewees said that it is insufficient. Large majorities in almost all of the countries surveyed mention television as the main source of information on nuclear energy. When assessing the information about energy and nuclear energy offered to children, EU citizens regard it as only slightly more adequate than the information in the media." [3]

9.2 Theory

From the PR point of view, basic knowledge of theses of communication science may always help to do a challenging communication job. That applies even more to issues related to nuclear power.

9.2.1 Watzlawick's Axiom 2

Each communication has both a content and an emotional aspect, the latter of which determines the first. "Each message not only has a specific content, but also an 'indication of how its sender wants it to be understood by the recipient." [4, p. 61] and [5, p. 482]. In other words, the aspect of content offers information, whereas the emotional aspect focuses on the relationship between sender and receiver. Therefore, public relations have to cope with the fact that purely informative communication does not exist. On the contrary, each expression contains a relationship statement. If a negative relationship is created on the content level, this can lead to a disturbed communication.

9.2.2 Gatekeeper Research

One research approach concerning public attention refers to the concrete role of the media. All efforts by PR officers seem pointless if the media do not address their issues and inputs. "Gatekeeper research" deals with the selection process by journalists. The core issue is the discussion of the limitation of the informational content by editors. They determine the topics that are considered relevant and thus are published. Thus, the editors also determine which events become public. This, in turn, influences the gatekeeper approach according to the image of reality that is generated by recipients. "In addition to the influence of individual predispositions on news selection, the influence of so-called 'institutional' factors was also recognized. This means that no journalist should be regarded as an isolated individual, but is always a member of a 'news bureaucracy' made up of various departments with different tasks." [5, p. 277]. This results in the specific relevance of journalist contacts for communication work, especially for crisis communication. A network to relevant editorial offices in general or individual key journalists in particular is the key to balanced reporting.

9.2.3 Agenda-Setting-Hypothesis

Agenda setting research also deals with the impact of media coverage on recipients. In 1963, Bernard C. Cohen first posed this thesis. "The core idea of this concept is the assumption that mass media do not influence what we think, but rather determine what we do have to think about. To a certain extent, they determine which topics we put on our agenda." [5, p. 248]. Hence, the media, by means of frequency, scope, and presentation, show the importance of an issue that the public is provided with. Thus, media take over the selection, bring up and weigh a subject for discussion, and thus create a social reality for the public. The role of the media is an essential aspect of a crisis.

9.2.4 News Value Theory

An important concept for the explanation of the news selection by the mass media is the "News Value Theory". "While classical gatekeeper research only observes the last stop on the route from the event to the editorial office, the News Value Theory starts much earlier in the perception of the events themselves." [5, p. 279]. The basic assumption of this approach, based on Walter Lippmann [6] and his term of "News Value," is that events have certain characteristics making them worthy of attention or interest. These properties or message factors determine the message value, that is to say the publication worthiness of an event: The more pronounced these properties are, the greater the message value of the event.

9.2.5 *RACE-Formula*

The RACE (Research-Action-Communication-Evaluation) model is a four-step process for communications planning proposed by John Marston [7]. Developed by Klimke and Schott [8], the RACE-formula compresses the steps necessary to overcome a crisis.

- *Research*: During the first phase of a crisis, there is sometimes a kind of shock and often there is only a few or non-reliable information available. In such a situation, it is all about clarifying the most important questions, which are when, what, who, where (how, why). This phase is crucial for the further development since no strategy can be defined without reliable information.
- *Action*: Based on the results gained through this research phase, one will be able to develop a strategy, define one's communication goals, and set appropriate tactics.
- *Communication*: The results of the first two phases again form the basis for the public relations work to be started now. In addition, the spokesperson has to play a key role therein. It is all about providing reliable information and being credible and trustworthy. In addition to the classical instruments such as press conferences and press releases, the spokesperson's performance significantly influences the perception.
- *Evaluation*: After the crisis, it is important to evaluate whether the chosen strategies, tactics, and measures have been chosen correctly. The objective in this context is learning for the future respectively being prepared for upcoming crises.

9.3 Practice: Case Studies

9.3.1 *Fukushima (Japan)*

On Friday March 11, 2011, the day when the disaster in Fukushima occurred, I had been going for a walk in my hometown Langau and received urgent phone calls from journalists from the Austrian Broadcasting Company (ORF) and the Austrian Newspaper *Der Standard.* Each of them wanted to talk to a nuclear expert about the accident and the current situation at the NPP Fukushima Daiichi. As the university's spokesperson, I had to rely on my univeristy-wide (internal) network to the relevant experts and could easily provide the media with reliable contacts at TU Wien. When the triple-disaster in Fukushima occurred, the colleagues at the Institute of Atomic and Subatomic Physics in Vienna (ATI) proved to be communication professionals. As a globally acting technical university with its focus on research, one of the duties of our scientists is to communicate and offer a rational evidence based expertise, especially in case of an emergency. When the disaster evoked, tremendous insecurity and fear all around the world and also among the people of Austria, they helped and transferred unbiased information. While universities usually complain about lack of funding and understaffing, the flexibility of a university with its student volunteers as a highly motivated human resource made it possible to quickly set up an "Information Center." Here, dozens of enthusiastic students, under the guidance of the nuclear experts at the ATI, researched for reliable information and answered urgent questions by the media and the concerned public via e-mail and phone [9]. The work of the Information Center made the ATI and TU Wien "the source point for the most recent and most reliable information available about the Fukushima nuclear accident in Austria" (ORF, Zeit im Bild 2, March 2011). Within weeks, the ATI and its Information Center provided expert information for 28 TV interviews, 26 radio interviews, and 52 print and online media interviews. Further, it provided unbiased and personalized information for approximately 900 inquiries by private individuals who had contacted the Information Center via e-mail or telephone [9]. While news organizations often struggle with a lack of manpower that would be needed upon a sudden crisis, a university—when endowed with a positive spirit—can rely on the availability of enthusiastic students who will be more competent in assessing the quality of potentially dubious raw data than journalists.

The experts at ATI took responsibility, referring to the university's mission statement "technology for people". Let me quote Prof. Georg Steinhauser (then at the ATI), who replied to a heckler during a public speech in the Austrian Federal Ministry of Health, why he would always emphasize his expression of hope when talking about the Fukushima accident and not embark on possible worst-case scenarios, "People are genuinely afraid of radiation already. I firmly believe that it is the obligation of science to allay this fear, not to fuel it." [10].

Subsequently, in September 2011, a large European joint project investigated the arrival and dispersion of the radioactive plume over Europe after the reactor accident in Fukushima [11]. The experts at the ATI of TU Wien also provided data and expertise to this joint study. As a research institution, the great strength of a university is its great degree of flexibility that allows the allocation resources to quickly establish international collaborations and research projects for the monitoring of airborne contaminants. Moreover, it is the "freedom of research" that allows participation in international large-scale projects, thus allowing the linking of measurement data from all over Europe, so that the spread of radioactive substances from Fukushima could be reconstructed exactly. The outcome of the study [11] was that the releases from Fukushima did not pose any health hazard to the European people at any time.

9.3.2 The TRIGA Mark II Research Reactor in Vienna (Austria)

In 2012, another phone call was the trigger for a challenging communication process. The call reached my colleague and reactor operator Dr. Mario Villa at the ATI. It was a call from the Department of Energy of the United States (DoE) asking,

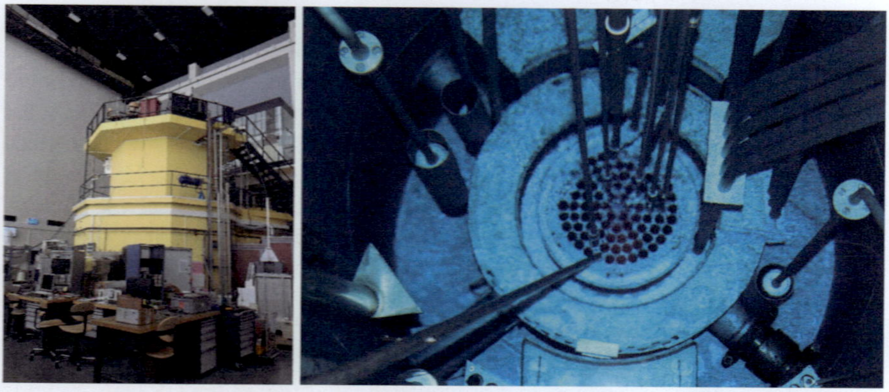

Fig. 9.1 The TRIGA Mark II reactor of TU Wien (left) and its core (viewed through the pool) at full power (250 kW) (right)

"Who is this person?" They were referring to media reports in which the university's spokesperson was quoted with information concerning the research reactor at TU Wien. At that time, the university had to cope with the exchange of the fuel elements from its TRIGA Mark II (Training, Research, Isotope Production, General Atomics) reactor. The university required the exchange of the fuel elements in its research reactor for being able to operate the infrastructure for 50 more years. After negotiations with the US Department of Energy and representatives of the Austrian Ministry of Science, the TU Wien agreed on the regular exchange of the fuel elements in the said research reactor and signed a corresponding agreement with the US Department of Energy and EURATOM. The agreement covered the withdrawal of 91 fuel elements and delivery of 77 low-enriched ones (Fig. 9.1).

At this point, let us take a look at the big picture. TU Wien is among the most successful technical universities in Europe. It is Austria's largest scientific-technical research and educational institution, and operates the only research reactor in Austria. Throughout its 200-year history, the TU Wien has always been a prestigious institution but, more importantly, also a modern research university with high aspirations. The mission statement of TU Wien is "technology for people." It expresses the high sense of responsibility of its researchers and teachers towards society. In 2012, one year after the Fukushima accident, there was still a high sensibility among Austria's society concerning nuclear energy and nuclear plants. It is precisely the already mentioned nuclear-critical attitude of Austrian people and subsequently of Austrian policy, which requires the corresponding expertise in the areas of nuclear reactors and radiation physics as well as radiation protection. The research reactor is the best possible training facility and is after the closure of the two other research reactors in Graz and Seibersdorf the only nuclear research center in Austria. As the disaster in Fukushima has clearly shown, it is indispensable, especially for nuclear emergencies, to possess national expertise. This concerns issues such as the emergence and geographical spread of radioactive radiation, food safety, limit values for humans, animals, the environment, and goods. Such knowledge is of particular importance in view of the proximity of nuclear power plants surrounding Austria.

It was a special situation in 2012, as TU Wien had to manage an additional issue. During that time, the university had an internal financial crisis. For that reason, the university faced the accusation of financing a reactor from NGOs and the political Green Party (Die Grünen), as if "Financing a reactor while having no money for students ..." etc. Therefore, what the management had to anticipate was the challenge of communication although the university had agreed with the USA and the Austrian authorities not to disclose any details due to security reasons. Nevertheless, it was necessary to inform the public and offer reliable data. The result of our activities in public relations and press review showed 29 media reports from January 2012 up to December 2012, which aimed to be objective. It is clear that the yellow press had the very alerting headlines such as "Nuclear Alarm in Vienna," for example. As I was responsible for PR and was the only one who talked to journalists, it was my name in the newspapers that provoked the question of the US DoE "Who is this person?".

9.3.3 Langau (Austria) and Dukovany (Czech Republic)

Referring to the results published in the repeatedly mentioned Special Eurobarometer 324 [3] on "Europeans and Nuclear Safety," I interviewed the mayor of my hometown Langau, Ing. Franz Linsbauer in May 2018. Langau (AT) is approximately 35 km linear distance to the town of Dukovany in Southern Moravia (CZ). NPP Dukovany was put into operation between 1985 and 1987 and consists of four reactor blocks with a total capacity of 1,792 MW. The existing plant comprises four WWER-440/213 Soviet-type pressurized water reactor units (4 × 440 MW electrical power, built 1979–1987), a spent nuclear fuel storage facility, and a storage for low and intermediate level radioactive waste. Up to two new reactor units are currently planned for the Dukovany NPP site, with an electrical output of up to 3500 MW. In July 2016, the Czech Ministry of Environment opened a cross-border procedure under the Espoo Convention and the EIA Directive on "New Nuclear Power Plant at Dukovany, Czech Republic."

The Special Eurobarometer 324 [3] dealt among other issues with topics such as "Knowledge of nuclear issues and nuclear safety; Information on nuclear energy and safety: people's feeling of being informed, whether there is sufficient information in the media and schools, preferred information sources; Decision-making and participation: the level of decision-making and willingness to participate." Within this context, I asked Langau's Mayor Linsbauer about his assessment in his capacity both as a representative of local authority and as an organic farmer in a borderland municipality. "From my point of view, it is a very frightening situation for us all. On the one hand, we have the information that NPP Dukovany is no longer a modern power plant; on the other hand, the additional new reactor units will strengthen the site and so our children are extremely unlikely to experience decommissioning of Dukovany. It appears to me that in the event of an expansion any chances for such decommissioning in the foreseeable future will be lost."

In reply to the question whether or not, from his perspective, the comprehensibility and transparency around NPP-projects near the border are sufficient, Linsbauer said, "The only useful information is the one coming from the federal State of Lower Austria. Personally, I would expect more involvement of the Federal Ministry for Sustainability and Tourism. Of course, there are offers like the participation in "Klimabündnisgemeinde" (Climate Alliance Municipality), "Bodenbündnisgemeinde" (Soil Alliance Municipality) or established "model regions," but the focus of the latter is different (e.g. charging stations, photovoltaic systems, and thermal insulation). Fundamentally, important things remain rather unaddressed. In Langau, a very small community with 670 main residents, we have gained 370 signatures against a planned near-border nuclear waste repository. That shows that we were able to mobilize more than half of our population for this issue. However, the campaign mainly based on the initiative of the Government of Lower Austria, especially on Governor Mag. Johanna Mikl-Leitner and Commissioner of Environmental Affairs Dr. Stephan Pernkopf. Additional information about nuclear issues comes through initiatives such as "Energiestammtisch" ("Get together on Energy") or dedicated individuals who disseminate information. Me, as Mayor and farmer, I have a strong network with people in the Czech Republic, such as politicians, farmers, hunters, and other citizens. Therefore I am quite aware that the Czechs have a completely different approach to nuclear energy." As there seems to be a lack of systematic exchange with the Czech authorities, it largely depends on the active management of personal networks. "Exactly," Linsbauer says, "my experience is that along the borderline, in small communities, there are far more nuclear energy proponents, while inland the voices become more critical. It is very rare that one finds a real nuclear opponent. Among my Czech hunting colleagues, I do not find an opponent. On the other hand, music colleagues from the Czech Republic e.g. in Dačice or the City of Brno get excited when they hear or talk about nuclear power."

9.3.3.1 Schools as Educational Platforms for Nuclear Questions

"EU citizens regard the information about energy offered by schools to children as only slightly more sufficient than that from the media. 58% of Europeans say that this information is not sufficient for children to acquire 'a basic knowledge on the risks and benefits of energy choices in general and nuclear energy in particular'. However, 29% think that this information is probably or certainly sufficient ('certainly', 5%, or 'probably', 24%). This last result has improved somewhat compared to three years ago (+4 points)" [3, p. 95]. Facing this apparent lack of "nuclear education" in the European school system, I asked mayor Linsbauer whether or not he was aware of any special trainings or information campaigning in his community's school or preschool related to possible incidents in the nuclear power plants across the border. "In regular intervals, we carry out civil protection exercises aiming at providing guidance to teachers on how to act in case of an emergency. Primarily, we focus on fire hazards, but of course, other incidents such as flood events or nuclear disasters are also relevant. This activity forms part of the disaster protection plan that each community must have. We have organized a corresponding core team representing the commander of the voluntary fire brigade, a resident paramedic, the civil protection officer and community representatives such as mayor and vice mayor. In the case of Langau, the comradeship alliance's chairperson employed with the Austrian army is involved as well. We do not do NBC (nuclear, biological, and chemical) protection exercises, yet. But together with our fire department we train various scenarios such as rescuing pupils through the window and showing them the meeting at the collection point in the municipal office." Does this mean that the community Langau is prepared? "We do our best! Realistically I have to admit that for a nuclear accident we are not sufficiently prepared. Regarding our limited capacities, we only could accommodate less people than desirable. This includes e.g. a protection cellar. I could imagine launching a project in which we prepare the basement of our school building as a protection room. In case of an NPP accident, we could give shelter to 60–70 children and teachers. If the municipal had money left, I could imagine realizing this project. I think our conversation is even helpful, because one becomes aware of prevention needs again. It's no big deal organizing an exercise and I have the impetus to set up an emergency room."

Based on the considerations above, the civil protection plan is organizationally anchored and follows a clear structure. That brought me to the question whether or not there are special communication measures included in this plan. "The priority is life and limb. Moreover, it is part of our crisis policy to involve other municipal staff for instance the members responsible for Public Relations who edit the local newspaper 'WILLI' [12]. The online editorial office of Langau's website is also in the loop in order to react quickly. In addition, the fire brigade with their command vehicle, having appropriate radio- and loudspeaker equipment, can keep people informed by announcements, even when other channels should fail. This is pure information work. We want to inform people comprehensively so that they are enabled to assess the situation by themselves. In my opinion, such scenarios call for the truth. You do not have to gloss over anything. People need to know and hear when something has happened so that they can assess it correctly. It does not help just to say that there was only strong wind and in fact, it was a blast. One must not play down anything!"

9.3.3.2 The Role of Science in Nuclear Crisis

The scientists and experts dealing with nuclear technology are responsible to society with regard to questions such as how does the technology work? What are the hazards? What are the disadvantages? Are there any benefits? What are the alternatives? Asking Mayor Linsbauer how he sees the role of science in this context, he answered, "I think it is very important that scientists also point out what the benefit of nuclear energy is and also what are the alternatives. Facing our hunger for energy, I can hardly imagine the meaning of transposing this consumption to the whole world. Let me give you an example. These days, there are billions of solid cubic meters of wood infested with the bark beetle. In the Czech Republic, they have no need for this wood. What happens to it? They export it to Austria for a dumping price and thus push prices down while we import nuclear power. That is what I am expecting from science, to tell that there are advantages of nuclear energy in terms of easing the burden on climate and environment, but who can tell us the risks and effects and predict what the burden will be, let's say, in 200 or 500 years?"

9.4 Conclusion

Decision-makers and managers are well-advised to prepare themselves together with an experienced, responsible person or team. Concerning the organization of public relations in case of an emergency or a critical situation, the first question is "Who is the person in charge of communication?". They must find a qualified person and charge it with the task. And it must be ensured that such a person in charge taking care of this communication agenda is reliably available in the event of a problem. They are encouraged to inform this person from the very beginning and involve this person. The credo is "Everything we communicate must be true, but on the other hand, not not every single aspect of such truth needs be communicated automatically," especially when the information is capable to increase uncertainty or confusion. A lesson to be learned from the TRIGA'S refueling procedure is that if somebody comes up with the question "Who is this person?" the communication procedure worked efficiently and precisely. In fact, the question indicated that the PR had been organized professionally. This offers the opportunity of controlling the information and communication processes. At management level, the most important question is "Who is the person designated as a first point of contact for the public?" In the words of Naoto Kan, former Prime Minister of Japan, "Retreat is not conceivable." Screening the media coverage around the Fukushima nuclear accident, this one sentence became central. It was a clear assignment to the power plant operators in Fukushima and as it was a promise for the Japanese population.

The strategy during the nuclear fuel exchange of the ATI's TRIGA Mark II research was as follows: By communicating the importance of the research that is being conducted using this nuclear facility, including facts and figures of the project, the university showed up with reliable information and gained high credibility. Facing a critical situation or a crisis, the responsible person for communication needs a clear-cut picture from the very beginning. As the refueling procedure was an international cross-border cooperation, it was necessary to have all the information and access to the operation plan from the very beginning. Therefore, a strategy for communicators and public relations officers is to get involved and to analyze the entire process by answering the questions "When? What? Who? Where?". It is crucial to embody the chains of command between experts, authorities, law enforcement officials, and other communicators. The communication processes must be defined precisely. Last but not least, it is up to the communicator to analyze all groups of relevant stakeholders, have recourse to the network of journalists and stakeholders, and to be on the spot.

In times of nuclear emergencies, building confidence seems to be the key issue in public relations and emergency communication. Confidence can be built by sticking to the truth and to facts, even if they appear inconvenient. There is no universal communication strategy for each and every case. However, my personal expericence was that "Truth is reasonable for mankind" (Ingeborg Bachmann). Providing public with reliable expertise and data is an effective means to reduce fear and to build trust. The handling of the Fukushima crisis by the nuclear experts at ATI illustrated the great strength of a university in providing information to the public by its flexibility and the availability of enthusiastic students who are willing to volunteer when it comes to solving a (nuclear) crisis. A university can develop great strength in crisis communication and become a source of trustworthy and unbiased information to the public and the media.

References

1. Austrian Federal Law of 15 December 1978 on prohibition of exploitation of nuclear fission energy for energy production in Austria. Federal Law Gazette No 1978/676; 1978.
2. Special Eurobarometer 297. European Commission. 2008. http://ec.europa.eu/commfrontoffice/publicopinion/archives/ebs/ebs_297_en.pdf. Accessed Jan 2019.
3. Special Eurobarometer 324. European Commission. 2010. http://ec.europa.eu/commfrontoffice/publicopinion/archives/ebs/ebs_324_en.pdf. Accessed Jan 2019.
4. Watzlawick P, Beavin J, Jackson D. Menschliche Kommunikation. Bern: Hogrefe; 2017.
5. Burkart R. Kommunikationswissenschaft. Vienna: Böhlau UTB; 2002.
6. Lippmann W. Public opinion. New York: Harcourt Brace and Co; 1922.
7. Marston J. The nature of public relations. New York: McGraw-Hill; 1963.
8. Klimke R, Schott B. Die Kunst der Krisen-PR. Paderborn: Junfermann; 1993.
9. Steinhauser G, Villa M, Bernt N, et al. Information management of the Fukushima reactor accident in Austria. Disaster Adv. 2012;5(2):61–3.
10. Steinhauser G. Tschernobyl—Fukushima. Ein Vergleich der zwei schwersten Reaktorunfälle. Public lecture (in German) given at the Festival Room of the Austrian Federal Ministry of Health, January 8, 2014, Vienna, Austria; 2014.
11. Masson O, Baeza A, Bieringer J, et al. Tracking of airborne radionuclides from the damaged Fukushima Dai-ichi nuclear reactors by European networks. Env Sci Technol. 2011;45:7670–7.
12. WILLI. 2018. http://www.langau.at/willi.html. Accessed Jan 2019.

Chapter 10
"Fukushima Live": About the Role and Responsibility of the Media

Sven Stockrahm

Abstract Competent and independent (science) journalism is needed not only, but especially in times of crisis. The accident at the Fukushima Daiichi nuclear power plant in March 2011 can be viewed as an example of the role and responsibility the professional media has to take on in the age of the digital revolution.

A limited empirical evaluation of German news media coverage during the nuclear emergency response serves as a case study here. By reflecting on decision-making processes at ZEIT ONLINE, being one of the largest platforms for online journalism in the German-speaking countries, five paradoxes and implications are discussed.

Today the ability to report immediately includes the obligation to do so, with all its shortcomings. Journalists have to precisely tell users, readers, and viewers what they know and what they do not know as well as constantly correct mistakes and clarify circumstances.

Additionally, as debunking misinformation can easily strengthen the belief in false facts, accurate and transparent reporting without speculation is essential.

Science and facts can help people make informed decisions when presented in a way that embraces unknowns and explains behavioral patterns. But science cannot only prove a point but also be used to manipulate truth. Therefore, journalists and scientists need to work together—respecting a professional distance—to value their shared asset of credibility.

To report facts truthfully and to report the truth about a fact is a process that has to be refined and adapted constantly.

Sven Stockrahm is a science journalist, editor, podcaster and staff writer for the German news site ZEIT ONLINE. Since 2018, he is the deputy head of the science, technology, and digital news section. His reporting on the Fukushima Daiichi nuclear accident has been recognized as outstanding achievement by the jury of the German Axel-Springer-Prize in 2012.

S. Stockrahm (✉)
Science and Digital News, ZEIT ONLINE, Berlin, Germany
e-mail: sven.stockrahm@zeit.de

G. Steinhauser et al. (eds.), *Nuclear Emergencies*,
Current Topics in Environmental Health and Preventive Medicine,
https://doi.org/10.1007/978-981-13-8327-4_10

Keywords Science journalism · Journalism · Reporting · News judgment · News · Media · Fact checking · Public interest · Storytelling · German news media · Online journalism · Fukushima nuclear accident · Germany · Japan

"It is no longer enough to report the fact truthfully. It is now necessary to report the truth about the fact." This quote from the 1947 report of the Commission on Freedom of the Press, better known as the Hutchins Commission [1], might be more important than ever in today's journalism and mass media. Moreover, it can be viewed as a constant reminder of the importance of competent journalism in the unfolding and aftermath of major incidents such as the Fukushima Daiichi nuclear accident.

To explain why, one has to look back to the time right after World War II. When the Hutchins Commission report set out to answer the question, "Is the freedom of the press in danger?" the press could easily be used by only a few in power as an "instrument of mass communication" to control and manipulate the public. It had in fact "greatly decreased the proportion of the people who can express their opinions and ideas through the press."

Today, this has changed dramatically. In the age of the digital revolution, the possibilities to express one's own views have never been more abundant. Communication tools like smartphones, tablets, and desktop computers are now at the hands of billions of people worldwide. They allow for a constant stream of information that is in the now, online, 24/7. This might have blurred the line between private opinion, corporate communication, advertising, and journalism. It has become difficult to discern between facts, opinion, and the sources of information.

To report the truth about the fact means, therefore, more than ever to be transparent about where data and news have originated from. Who is providing information, by which means, and for what reason? Is the news coherent, does it make sense, is the source itself trustworthy?

All these questions surrounded the media coverage about Fukushima. They are critical in order to understand why independent journalism is needed not only in time of crisis.

But what does competence in journalism entail?

In the following, I discuss the role and responsibility of the media using Fukushima as an example. Intentionally, this is a personal empirical account and evaluation. It is by no means a complete assessment, but a structured collection of thoughts and conclusions. The focus will be on media coverage in and from Germany and not from abroad or within Japan. I am not a scientist but a graduated science journalist. I have been working for the German news media ZEIT ONLINE and DIE ZEIT as an editor and writer since 2008. I am currently the deputy head of the science, technology, and digital news section at ZEIT ONLINE.

ZEIT ONLINE is one of the largest and fastest growing platforms for online journalism in the German-speaking countries. It stands for outstanding multimedia journalism, award-winning storytelling, and data journalism. Its editorial team covers breaking news, publishes analyses, reviews, and opinion pieces on politics, business, science, culture, and society. These reach almost 15 million (unique) users monthly [2].

At the time of the Fukushima nuclear accident in 2011, I was responsible for reporting about most of the events within the first hours and days as well as in the weeks and months after the magnitude 9 earthquake and the devastating tsunami hit northeastern Japan. As a reporter, I visited the Tohoku coastal region that was hit the most 6 months after the quake.

Working for a news website means that you do not have to wait for printing deadlines or even for setting up a live feed to a reporter. During a breaking news situation, we have to work in the now, we are constantly online, as are our readers.

In that respect, Fukushima can be viewed as an event that reflects certain paradoxes journalists and especially science journalists are dealing with and often times are criticized for [3]. How do we report news and when?

10.1 Journalism, Especially Online Journalism, Cannot Not Communicate

When the tsunami hit Japan on March 11th, 2011, ZEIT ONLINE set up a live news blog for days. Editors gathered around and tried to verify any piece of information that was sent by news agencies and news outlets from Japan. In the beginning and only for a couple of hours, the news mainly dealt with the devastation the tsunami had left behind. Then the attention shifted to the nuclear power plant in Fukushima. The defining event became not the natural disaster but the nuclear emergency.

The biggest problem at the time was that nobody knew exactly what was going on. Information provided by officials and the owner of the plant, Tepco, were inconclusive, contradictory, or not given at all. Evacuation orders and learning about the complete loss of electricity in the plant let everyone fear for the worst. This is when speculations began.

Major questions the news media were facing at that point referred to providing context for a situation that was still unclear and somewhat unpredictable. At ZEIT ONLINE we discussed to take a breath, to pause in order to not produce any (more) speculations but to stick to the facts and the unfolding event as it happened. We asked ourselves: Should we not wait until more information is available? Until official statements are released and the situation can be assessed a bit better?

Readers and viewers, who often ask such questions, for example, via Facebook, Twitter, or the comment section at ZEIT ONLINE, are right. Often the simplest context clarifies only days after an event, the more complex sometimes take years or decades to be fully understood.

So should we slow down? The answer is no. We cannot restrain ourselves while other media and people on the social web are reporting what they see and experience and do speculate.

The ability to report immediately includes the obligation to do so.

10.2 As Journalists, We Are Part of the Frenzy We Try to Fight

But the defining questions are how and what to report. Competent journalism provides information that is in the public interest. At its core it tries to figure out what happened, when, where, why and how, by whom, and with who involved [4].

During the first couple of days of the nuclear emergency, we decided to do two things: keeping the news blog with all its shortcomings while providing facts and knowledge about radiation and nuclear power plants, emergency planning, and all that.

Shortcomings included unconfirmed reports about what exactly was happening at the nuclear plant and information released by official news agencies, non-governmental organizations, ministries, public officials, and the company in charge of the nuclear plant. This accumulated news that often seemed contradictory or incomplete even flawed by news agencies and reporters trying to give context about a topic most of them had never before reported on. Radiation, radioecology, and nuclear emergencies involve a lot of science and details that can be difficult to explain or even understand. Challenges I myself had to deal with—I was basically thrown into the cold water as almost anybody else in newsrooms across the world.

I wrote an article that tried to lay out best and worst-case scenarios, asking physicists, physicians, radioecologists, radiation protection experts, and other researchers for their assessment. It was published only a week after the tsunami had hit Japan [5]. In doing so I was as well trying to make sure the experts were neither lobbying for nor against nuclear power, for example. But also they should make themselves clear about what could possibly happen and what not. Providing context being realistic and trying not to nurture hysteria.

As journalists, our task is to inform and explain and assess news. But Fukushima was difficult. Misinformation was spread, Tepco withheld data and it was just not clear even in the weeks and months after the initial accident, which source you could trust and which not. One of the biggest topics already in the first weeks after the nuclear accident was the fact that Japanese officials had obviously turned a blind eye or two in the past when it came to regulations for nuclear power plants. There has always been a very friendly and cooperative relationship with the nuclear industry—to the extent of cutting back on regulations and safety standards [6].

As reporters, we have to address all this. We have to precisely tell our readers what we know and what we do not know. And most importantly: We have to constantly correct mistakes and clarify circumstances.

10.3 Debunking Misinformation Can Strengthen the Belief in False News

"At the heart of journalism remains the neutral, unbiased report, still grounded in the traditional questions of who, what, when, where, why, and how," states the American writer, editor, and teacher of writing, Roy Peter Clark [7]. The press and competent journalism have become more important in a world that is questioning its role and demanding objectivity. In a world where everyone with access to the internet and a mobile device can go on air and report, journalism has not become obsolete, but essential again.

But journalism in its core has never been neutral or unbiased or even objective. Every editor, writer, and reporter knows or might guess this every once in a while doing the job. Journalism is about selecting news, fact-checking information, presenting it to support a certain truth. This process in itself is what makes journalism subjective to what the people who work in this profession have agreed upon. Journalism follows a purpose, has ethical standards, and relies on news judgment, literacy, evidence, and critical thinking. It reflects the culture we live in.

Journalism is the attempt of relieving one's own point-of-view from bias. A way to do this is by being transparent about what you choose to include in a news story and what you leave out. What information is assessed to be essential and which not?

In a breaking news situation, the assessment might be flawed. Especially when the situation reporters are dealing with, is unprecedented. This is what it was like in the wake of the Fukushima Daiichi nuclear accident.

Instead of choosing what to report, and addressing the lack of information, media outlets all over the world, and especially in Germany, lost their ability to clarify matters and succumbed to speculation.

During the first days of the Fukushima crisis, reporters on German television openly discussed the possibility of thousands of people dying as a result of the incident. News about possible fall-out over Japan's capital Tokyo had become a major topic, a lot of countries had advised their people to immediately leave the country and prepare for the worst. A lot of this was complete fear mongering and uninformed speculation.

Having witnessed three huge explosions on the site of the nuclear plant, anything seemed possible. It did not help much to constantly tell people that these explosions were caused by hydrogen and that the reactors themselves might still be intact. It just did not seem to matter anymore. The power of images had taken over; media outlets worldwide were showing the exploding reactor buildings on a permanent loop. They are now part of the collective memory of the Fukushima Daiichi nuclear accident. This almost destroyed any belief in the public eye that this emergency could ever be contained.

In some ways, this has not stopped in the years since the accident. There is still a lack of trust in the media coverage or the information released by official reports from the World Health Organization or the United Nations Scientific Committee on the Effects of Atomic Radiation about the possible health implications of the accident [8, 9]. This has to do with the fact that many people are scared of radiation and do not understand it well enough. It cannot be seen, felt, smelt, or tasted. In terms of a health threat, this can be seen as one of the worst dangers one might possibly deal with. Additionally, since at least the accident at the Chernobyl power plant in April 1986, nuclear emergencies are connected with misinformation, cover-up, and lack of communication [10].

Therefore it still is incomprehensible for a large part of the public that the Fukushima nuclear accident did not cause widespread radiation-related illnesses and that experts and health professionals do not expect dramatic increases in the number of cancers and other diseases in the future [9].

Another aspect to curb fears was the fact that the well-intentioned screening program for thyroid cancer in the aftermath of the accident was used by activists as proof of high radiation doses among the general public [10]. The screening of hundreds of thousands of children and young adults, in fact, found abnormalities and even cancer. It was hard to believe that these were due to overdiagnosis and overtreatment. And thyroid cancer specialists are still having a hard time to convince people that these abnormalities are in part harmless and would have remained undetected if it had not been for the screening program. It also did not help much to explain that thyroid cancer takes longer to develop and cannot be detected as a cause of radiation exposure in the immediate aftermath of a nuclear accident. And most importantly that if diagnosed, it can be treated fairly well [10].

Recent research has shown that trying to debunk misinformation can be very hard. In fact people who are presented with fabricated content that is presented in a factual way tend to believe in the accuracy of that information even more so if it is repeatedly shown to them. Tagging such stories as "fake news" and dispute them does not help. It might even increase their believability [11].

This phenomenon has been observed especially when it comes to social media platforms where inaccurate or blatantly false information can spread easily by sharing links to videos, stories, or images. Even more alarming are findings that debunking false information, meaning to present a corrective message that establishes that the prior message was misinformation, could as well strengthen the believability of the fabricated news [12].

This means that accurate and transparent reporting without speculation is essential.

10.4 Facts and Science Can Help People Understand and Make Rational Decisions

The less a danger can be assessed, the more powerful is fear and the feeling of being out of control. In the days following the nuclear accident, people were afraid and they doubted a lot of the news after the accident.

Especially in Germany with its long tradition of anti-nuclear protests, nuclear power and its effects have been considered particularly dangerous for decades. These fears were fueled even further when on 26 April 1986, an explosion and fires at the Chernobyl nuclear plant in Ukraine caused the largest uncontrolled radioactive release in the history of the civil nuclear industry [13].

Memories of safety measures and the lack of information are still vivid in Germany. Many citizens still think about contaminated wild boars in Bavaria when hearing about Chernobyl. Some remember well how children were told not to play outside after more and more details about the fallout of Chernobyl came to light.

So after Fukushima, people who were already against nuclear power, felt reassured that this is a technology that cannot be controlled. And a majority agreed. This led to the government's decision to abandon nuclear power earlier than planned. Chancellor Angela Merkel said that Fukushima had shown that the risks of nuclear energy are uncontrollable [14, 15]. The decision was almost entirely based on emotions rather than on safety concerns. In fact, the safety standards for the use and generation of nuclear energy remained almost entirely unchanged after Fukushima. The technology itself had frankly become unacceptable. Three months after Japan's nuclear accident, the German Bundestag passed legislation to abandon the commercial use of nuclear power by 2022 [14, 15].

This illustrates another misperception. As journalists and experts we can try to explain over and over again that, for instance, it is, statistically speaking, much more likely to be hit by lightning or to die in a car accident than being affected by a nuclear emergency. But to do so is a fallacy from the pre-school of risk assessment. Risks cannot be reckoned up against each other so easily. Smokers are also allowed to be afraid of nuclear power plants even if they cause themselves much more harm by lighting their next cigarette.

Journalists should explain how people perceive risks and it can help to talk about why people overestimate certain risks. Unlike politicians often suspect, explaining risks does not make people more afraid. But refusing to give an answer at all contributes to uncertainty and fear.

As journalists, we should not only ask: How dangerous is something? But also: What are the arguments that nothing will happen to me?

People underestimate risks that occur continuously and over a long period of time; Those which affect many people at a specific time trigger an outcry. In 2017, 3177 people were killed in traffic accidents in Germany [16]. Would they all die on a single day—we would probably immediately stop driving cars.

10.5 Journalists and Scientists Have to Work Together and Stay Discerning

As journalists, we need to inform and assess. But we do not need reporters who speculate about the nuclear apocalypse without any knowledge base. There are already enough people on the social webs who do so.

But what journalists can and should do, is make expertise heard. Especially science journalists need to remain close to the scientific community. This includes verifying the reliability of the experts they are talking to and choose to include in a story.

It is important for journalists to report on research and use science and scientific approaches to interpret the world as it is and as it may become. Journalists should be aware that science is not about irrevocable truths, but the process of getting to the bottom of what we perceive as factual and given. Science can answer questions, deliver context and causality, and clarify matters and situations.

There is no better way than to explain to people over and over again, what research into nuclear emergencies tells us and what not. As science journalists, we also have to report on the forgotten dangers and health issues after Chernobyl and Fukushima. Journalism has the means and the obligation to inform the public not only about inconvenient truths but also about topics that have been neglected.

This includes in respect to nuclear emergencies, for example, the psychological effects and the social impact of accidents. In Fukushima we see people having lost loved ones, their homes, and jobs, people who suffer from anxiety disorders, depression, trauma, and stigma. The fear of radiation could lead to long-term health effects that by far exceed the dangers of the contamination itself [17].

Keeping scientists and experts as sources, journalists should also never be afraid to call out misconduct, conflict of interests and question methods, and data that are presented. Science can be a tool to prove a point or a way to manipulate truth as well.

Scientists and journalists share a common asset that is essential for their expertise: credibility. It is as valuable as it is fragile.

To report facts truthfully and to report the truth about a fact or piece of information is also a process that has to be refined and adapted to. It includes certain elements, as the Poynter Institute, a non-profit organization for the advancement of journalism has laid out [7]:

- *News judgment*: What is important and interesting enough to be reported?
- *Reporting and evidence*: The gathering of reliable sources, the verification of information, and the distribution of evidence—the process and products of research.
- *Language and storytelling*: The effective use of language to express reports, stories, and other appropriate forms of communication.
- *Analysis and interpretation*: Critical thinking as well as context, meaning, trends, relationships, and tensions of information should be at the hands of a competent journalist.
- *Numeracy:* The ability to use computation skills and to report for numbers, to understand probability and statistics.
- *Technology*: Abilities in word processing, search and research functions, social networking, blogging, programming, mobile applications, data analysis and display, aggregation and curation of information.
- *Audio-visual literacy:* This is expressed, e.g., through photography and video, design and illustration, multimedia productions, the use of sound.
- *Civic literacy:* This includes knowledge of government, politics, social capital, social contracts, power, history, public life, civic culture, how audiences can be measured for public opinion, how media influence the constituent groups in society.

- *Cultural literacy*: This requires knowledge of and sensitivity to cultural differences, whether they are expressed by race, social class, ethnicity, religion, gender, or sexual orientation.
- *Mission and purpose*: These include media ethics and law, the history of journalism, principles of democracy, and a working knowledge of the role journalism plays in communities and municipalities.

Does being persistent in reporting facts and news in such a way work out? I personally believe so. The more time is passing since the Fukushima Daiichi nuclear accident, the public can see that Japan has not turned into a nuclear wasteland. The construction and decontamination work is perceived so much as a routine already that people, especially in Germany, have lost interest in reports about Fukushima.

This is not necessarily a good thing, but it shows that the attention span even for nuclear emergencies in the public eye is getting shorter and shorter. So not only as journalists do we have to make sure to continue to report. Scientists need to keep us informed, too.

References

1. The Commission on Freedom of the Press. A free and responsible press. The University Of Chicago Press, Chicago. 1947, p. 22. The Internet Archive. https://archive.org/details/freeandresponsib029216mbp/page/n37. Accessed 30 Aug 2018.
2. iq media marketing gmbh. https://www.iqm.de/digital/marken/zeit-online/. Accessed 11 June 2019.
3. Wegner J. Die fünf Paradoxien der Livemedien und der Mythos des Oknos. ZEIT ONLINE. 2016. https://www.zeit.de/gesellschaft/2016-07/online-journalismus-medien-amoklauf-muenchen. Accessed 21 Aug 2018.
4. Kovach B, Rosenstiel T. The elements of journalism: what newspeople should know and the public should expect. New York: Random House; 2014.
5. Stockrahm S. Von Super-GAU bis glimpflich ist alles möglich. ZEIT ONLINE. 2011. https://www.zeit.de/wissen/umwelt/2011-03/fukushima-szenarien-japan. Accessed 20 Sept 2018.
6. Stockrahm S. Japan drückt sich um eine Perspektive nach Fukushima. ZEIT ONLINE. 2012. https://www.zeit.de/wissen/umwelt/2012-07/fukushima-abschlussbericht-zukunft-kommentar. Accessed 20 Sept 2018.
7. Peter Clark R. The pyramid of journalism competence: what journalists need to know. Pointer. 2014. https://www.poynter.org/news/pyramid-journalism-competence-what-journalists-need-know. Accessed 28 Sept 2018.
8. World Health Organization. Health risk assessment from the nuclear accident after the 2011 Great East Japan earthquake and tsunami, based on a preliminary dose estimation. Geneva: WHO; 2013.
9. United Nations Scientific Committee on the Effects of Atomic Radiation. Levels and effects of radiation exposure due to the nuclear accident after the 2011 great east-Japan earthquake and tsunami. Volume I scientific annex A. New York: United Nations; 2014.
10. Stockrahm S. Die Strahlung ist nicht das Schlimmste. ZEIT ONLINE. 2013. https://www.zeit.de/wissen/umwelt/2013-02/fukushima-langzeitfolgen-radioaktivitaet-psyche-tschernobyl. Accessed 28 Sept 2018.
11. Pennycook G, Cannon T, Rand D. Prior exposure increases perceived accuracy of fake news. J Exp Psychol Gen. 2018. https://doi.org/10.2139/ssrn.2958246.

12. Chan M, Jones C, Jamieson K, Albarracín D. Debunking: a meta-analysis of the psychological efficacy of messages countering misinformation. Psychol Sci. 2017;28(11):1531–46. https://doi.org/10.1177/0956797617714579.
13. World Health Organization. Chernobyl at 25th anniversary – frequently asked questions – April 2011. WHO, Geneva. 2011. http://www.who.int/ionizing_radiation/chernobyl/20110423_FAQs_Chernobyl.pdf. Accessed 28 Sept 2018.
14. Deutscher Bundestag. Atomausstieg und Energiewende als "Herkulesaufgabe". 2011. https://www.bundestag.de/dokumente/textarchiv/2011/34716466_kw23_de_atomgesetz/205630. Accessed 28 Sept 2018.
15. Deutscher Bundestag. Die Beschlüsse des Bundestages am 30. Juni und 1. Juli. 2011. https://www.bundestag.de/dokumente/textarchiv/2011/34915890_kw26_angenommen_abgelehnt/205788. Accessed 28 Sept 2018.
16. Statistisches Bundesamt. Pressemitteilung Nr. 063 0,9 % weniger Verkehrstote im Jahr 2017. 2018. https://www.destatis.de/DE/PresseService/Presse/Pressemitteilungen/2018/02/PD18_063_46241.html. Accessed 28 Sept 2018.
17. Maeda M, Oe M. Mental health consequences and social issues after the Fukushima disaster. Asia Pacific Journal of Public Health. 2017;29:36S. https://doi.org/10.1177/1010539516689695.

Chapter 11
Teaching Radioactivity: What Is the Goal of Education?

Katsumi Shozugawa

Abstract Lectures on radiation are held at any university. In Japan, new radiation education became necessary after the Fukushima Daiichi nuclear power plant accident. According to our survey, the ratio of the student who eats/buys apples, which contained sufficiently lower than the Japanese regulatory limit, is approximately half. This chapter reported the status of measurement and understanding after 7 years of the Fukushima accident.

Keywords Fukushima Daiichi nuclear accident · Teaching · Public education Public perception · Radiation · Radioactivity

11.1 "2 Bq/kg of Apples"

"I offer you these apples that were grown in Fukushima Prefecture, and they contain 2 Bq/kg of radiocesium" and "I hope that you will discuss this with your family and friends based on the new knowledge you learned."

This was an inquiry for the freshmen and sophomores at University of Tokyo after they had attended our lecture on radiation, "Scientific understanding of radiation" (see Fig. 11.1). This lecture is held every year with the cooperation of many faculty members with an aim to disseminate the basic knowledge on the Fukushima nuclear accident that occurred in the spring of 2011. Most of the students, who attended this lecture, had been in junior high school at the time of Fukushima nuclear accident, and their initial response regarding radioactivity is best summed up with, "I do not understand it, but it is always on the news."

Our lecture included not only the commentary and definitions provided in conventional textbooks but also the real-life aspects of the issues [1]. Regarding radiocesium in Japanese food, the lecture has been developed so that they are organically

K. Shozugawa (✉)
Graduate School of Arts and Sciences, The University of Tokyo, Tokyo, Japan
e-mail: cshozu@mail.ecc.u-tokyo.ac.jp

G. Steinhauser et al. (eds.), *Nuclear Emergencies*,
Current Topics in Environmental Health and Preventive Medicine,
https://doi.org/10.1007/978-981-13-8327-4_11

Fig. 11.1 "Scientific understanding of radiation" in University of Tokyo

connected to knowledge obtained from scientific viewpoints such as the containment and control of radioactive materials immediately after the accident to the present, the regulations that affect the public and safe limits of radiation exposure.

I told them, "Although the number of research institutes capable of measuring radiocesium at concentrations up to a few Bq/kg was limited before the Fukushima accident, the rapid spread of the measuring instruments has made it possible to detect low-level radiocesium concentrations below 10 Bq/kg." and "Apples cultivated in Fukushima city contain about 2 Bq/kg radiocesium."

Keeping these scientific facts in mind, I suggested the idea of framing a discussion between food producers in Fukushima and consumers. Two opinions were given in response. One side of the argument can be summarized as, "Since the Japanese regulation for general food is 100 Bq/kg radiocesium, 2 Bq/kg of radiocesium in food is only 2% of that regulatory limit. Even if I eat an apple, the additional exposure dose is extremely small (0.007 μSv), which is negligible for human health. Thus, I will buy and eat them."

The counterargument is typically expressed as, "We have another choice, i.e., there are apples in supermarket that are cultivated in some other parts of Fukushima, where the radiocesium content is much lower than 2 Bq/kg. In addition, you are telling me that the apples have 2 Bq/kg of radiocesium, but I have never seen such an indication in the supermarket. So, I will not buy and eat them."

Figure 11.2 shows the percentage of positive (i.e., "I will buy and eat the apples": A) and the negative response ("I will not buy nor will I eat the apples": B). These results were collected using an online questionnaire that was sent to the students who attended my lecture. The total number of students who responded to the questionnaire was 42, and the percentage was 52% (B: will not buy) vs 48% (A: will

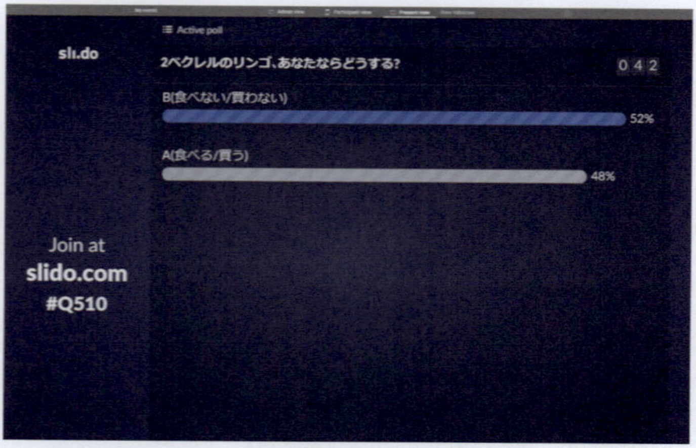

Fig. 11.2 The percentage of my question "Do you eat/buy 2 Bq/kg of apples?" A is YES (48%), B is NO (52%), respectively

buy). Interestingly, the percentage is consistently evenly split in lectures at the University of Tokyo and at other universities.

From an objective and scientific perspective, opinion A is correct. However, in my personal opinion, both opinions should be respected because the lecture audience judge beyond science after learning about radiation scientifically. Subjectively speaking, whether or not a person will purchase or eat the apples is not a simple matter that science can solve.

Ideally, the environment in which people obtain their scientific knowledge regarding radiation should be irrelevant to any other nuclear accidents. Among the people still living or considering to return to areas affected by the Fukushima nuclear accident, many refugees continue to suffer from several problems since the date of the accident, including evacuation, decontamination, and compensation. To ask these people to fully understand radioactivity is an added burden that is difficult to justify.

For that reason, sometimes, the situation gets worse because the complex radioactivity science confuses many people. At the same time, it has been my experience in the past 7 years that new radiation regulations that were forced after the Fukushima accident fall short of solving various problems with respect to radioactive materials.

Since the Fukushima nuclear disaster in 2011, I have presented radiation seminars to more than 8100 people, including public schools in Fukushima Prefecture. Along the way, I have realized that a complete understanding of radioactivity that includes recognizing the specific background of each situation is not adequately addressed by government regulations. For instance, huge costs and consciousness associated with decontamination have a dramatic effect on whether or not decontamination is even possible.

11.2 Measuring Food Is Essential in the Future

Seven years after the Fukushima nuclear disaster, the levels of radioactive contamination in the food sold in the Fukushima are far below the Japanese regulatory limit. A comprehensive radioactive monitoring system established by the government has guaranteed very low-level contamination. More specifically, we have assessed the radioactivity of more than 300 high-sensitivity samples, and even the milk samples sold in Fukushima displayed concentrations of only 0.094–0.33 Bq/kg, which was 1/150 of the regulatory limit [2].

Furthermore, I would like to introduce an example of agricultural crops in Iwaki city in the southern part of Fukushima Prefecture [3]. Regarding agricultural crops for shipment produced in Iwaki city, the number of samples measured by Iwaki city from September 20, 2011, to August 31, 2016, was 32,247. Among these, 36 products exceeded the regulatory limit (100 Bq/kg), which is about 0.1%. The percentage of crops displaying concentrations below the detection limit (10 Bq/kg; up to March 2012, it was 20 Bq/kg) accounted for 96.6% of the total. It was a clear fact that the number of samples that exceeded the Japanese regulatory in agricultural crops in supermarket was extremely low.

However, it should be noted that the radioactivity listed here was "for shipping" agricultural products, that is, it was limited to agricultural crops that were on the market. According to other statistics of measurement, "for self-consumption" and not aimed at shipping and selling, which was a measurement separated from "for shipping," the percentage exceeding the regulatory standard in Iwaki city was approximately 9.4% (28,099 out of 2653 crops exceeded the limit). Among these crops, 766 out of 1001 samples of Shiitake mushrooms exceeded the regulatory limit at a rate of 76.5%. Additionally, the rate of exceeding the regulatory limit was 95.0% for Koutake (kind of mushroom) and 5.4% for Yuzu (citron).

The amount of regulatory crops that exceed the regulatory limit has been kept very low because a stringent system for clearly classifying production methods between shipping and self-consumption has been adopted, and the current measurement system in Fukushima has guaranteed the low-level contamination.

Keeping this in mind, it is important to consider whether this ratio would increase if deregulation occurred on the grounds that the radioactivity in shipped crops was characterized with low-level contamination.

This is not simply a statistical problem. From the consumer's point of view, if crops exceeding the limit are detected, it could lead to a decline in the desire to purchase and/or trust in the reliability of the current monitoring system.

Since the half-life of ^{137}Cs (one of radiocesium) is 30.1 years, this problem is expected to continue for many generations. One of the effective countermeasures is to continuously keep on investigating a method that can overcome this. Instead, of taking measurements only at the "exit," measurement of actual conditions is important.

11.3 The Existence of Outliers

It may be surprising that the radioactive contamination of fish has been much lower than that of crops. Nonetheless, the affected Japanese fishery industry has implemented a voluntary regulatory limit at the 50% level of the governmental regulatory limit to exemplify the very high level of food safety. The Onahama Fishery Cooperative in Iwaki is pursuing their own agenda in terms of education of the public by allowing some transparency in the data acquisition and even the handling of samples (at least in terms of accessibility of the laboratory). Figure 11.3 shows the measurement laboratory of the Onahama Fishery Cooperative.

For Iwaki-landing fishery products, the exceeding rate was estimated to be 0.03% (1 out of 3332 samples, from 21 May, 2012, to 31, August, 2016). Only one case was reported in 2013 where the regulatory limit was exceeded, and none have been reported in the last 3 years. This trend is same as that observed in other Japanese cities.

I would like to also introduce a very rare and specific case. Our research group investigated shortfin mako shark (*Isurus oxyrinchus*) fished in Numazu City, Shizuoka Prefecture, in 2016. The radioactive cesium in the edible part taken from that one was measured, and it was confirmed that 707 Bq/kg exceeded the regulatory by about 7 times. From the characteristics of the detected radiocesium (based on the $^{134}Cs/^{137}Cs$ radioactivity ratio), all the determined radiocesium was to be derived from Fukushima nuclear plant.

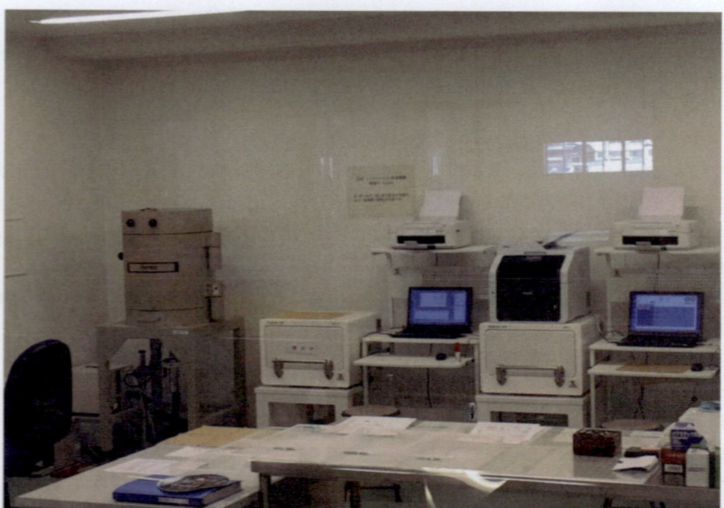

Fig. 11.3 Radioanalytical measurement laboratory of the Onahama Fishery Cooperative in Iwaki, Fukushima Prefecture

With the cooperation of the Fisheries Agency, Japan and Shizuoka Prefecture, tissues testing results of a shark captured off the coast of the Izu Peninsula, which was caught 500 km from the Fukushima nuclear plant, indicated that high levels of contamination were not present, not even close to Fukushima Prefecture.

Generally, in the case of large fish, radiocesium in their body ranges from 0.1 to 10 Bq/kg. According to Japanese database [4], a survey of the shortfin mako shark from April 9, 2012, to February 16, 2017, was conducted in Shizuoka, Tokyo, Miyagi, and Iwate prefectures. A total of 69 cases were reviewed, and the maximum radioactivity was 36 Bq/kg in the fish collected from the Miyagi prefecture. This statistic further proves that the high levels of contamination in fish are extremely rare.

The shark collected off Izu had seven times more radiocesium than the regulatory limit, which was an "outlier" even where radiocesium does not exceed the limit for almost all the fishery products. Since the consumption of single outliers is hardly dose-relevant, the Japanese government seems to promote the policy of terminating or deregulation of the monitoring system gradually. However, this leads to a decrease in the probability of finding outliers.

Owing to the small number of statistical outliers, the accumulated knowledge regarding the question of "why" is lacking. That is why the nationwide monitoring program for fish and crops should be maintained. Thus, it is important to pay attention to dynamics of radiocesium and the local and regional ecology over a period of time.

At the same time, what happens if outlier fish are widely reported under the current condition?

11.4 Lessons Learned in 7 Years After the Fukushima Accident

Seven years after the Fukushima accident, a lot of researchers have clarified the dynamics in terrestrial radiocesium behavior and migration including its presence in groundwater. However, enough knowledge has not been accumulated to apply this to all environments.

First, regarding agricultural crops, it will be necessary for the future to continue data collection using on-site measurements based on the future prediction using knowledge obtained in the past 7 years.

Second, regarding fishery products, despite there are rare exceptions, contamination levels have declined significantly. Fortunately, direct leakage from the Fukushima nuclear plant to the Pacific Ocean has become an insignificant factor for the Ocean. However, since 2015, ^{3}H and radiocesium have continued to leak not only to Ocean but also to the landside from the Fukushima power plant, and countermeasures to protect against its discharge have not been completed [5]. From this viewpoint, it is important to maintain a strict monitoring system in and around the Fukushima site.

The ideal solution would be to return all the released radioactive materials back to the fuel rods in the nuclear reactor, though this is unrealistic. Along with the scientific knowledge obtained from the Fukushima accident, it is important to develop flexible strategies to optimize solutions according to each region. Only considering whether or not it conforms to the criteria is overly simplistic and will not solve the problem over the long term. I think 7 years should be enough time for the implementation of "lessons" for the revival of Fukushima. Education of students and the public will play an important role in this revival procedure, which will require long-lasting and continuing efforts.

References

1. Shozugawa K. Radiation measurement for Fukushima revival (in Japanese). Trends Sci. 2017;22(4):40–3. https://doi.org/10.5363/tits.22.4_40.
2. Shozugawa K, Saito T, Hori M, Matsuo M. High-sensitivity analysis of radioactive cesium in Japanese foodstuffs: three years after the Fukushima accident. J Radioanal Nucl Chem. 2016;307(3):2117–22.
3. Iwaki City Office. Results of radioactive material of Iwaki's agricultural, forestry and fishery products (in Japanese). 2016.
4. National Institute of Health Sciences. Database of radioactive materials in food. http://www.radioactivity-db.info/. Accessed 15 June 2018.
5. Shozugawa K, Hori M. Variation of the radioactivity of ^{134}Cs and ^{137}Cs in flowing-well water and sump water, Okuma town after the Fukushima nuclear accident (in Japanese). J Hot Spring Sci. 2016;66(3):179–87.

Chapter 12
Agriculture in Fukushima: Radiocesium Contamination of Agricultural Products

Keitaro Tanoi, Naoto Nihei, and Martin O'Brien

Abstract In this study, we examine the radioactivity monitoring data derived from Fukushima prefecture. The Japanese government has established very strict limits for radiocesium in food since April 2012 (100 Bq/kg for general food). The Fukushima prefectural government inspected foods and found that most of the agricultural products in Fukushima did not contain contamination above the laboratory detection limit. Moreover, the radiocesium concentration in food samples decreased over time for the first 2–5 years, and very few samples of forest products, marine, and lake products exceeded the limit 5 years after the accident. Potassium fertilization of arable land has been the most effective countermeasure, and the effort to create crops with low levels of cesium is continuing.

Keywords Agriculture · Food safety · Fukushima Daiichi Nuclear Power Plant accident · Inspection · Limits · Monitoring · Radiocesium contamination

Abbreviations

FDNPP	Fukushima Daiichi Nuclear Power Plant
MAFF	Japanese Ministry of Agriculture, Forestry and Fisheries
MEXT	Japanese Ministry of Education, Culture, Sports, Science and Technology
MHLW	Japanese Ministry of Health, Labor and Welfare

K. Tanoi (✉) · N. Nihei · M. O'Brien
Graduate School of Agricultural and Life Sciences, The University of Tokyo, Tokyo, Japan
e-mail: uktanoi@g.ecc.u-tokyo.ac.jp

G. Steinhauser et al. (eds.), *Nuclear Emergencies*,
Current Topics in Environmental Health and Preventive Medicine,
https://doi.org/10.1007/978-981-13-8327-4_12

12.1 Introduction

12.1.1 *The Fukushima Daiichi Nuclear Power Plant (FDNPP) Accident*

The 2011 Tohoku earthquake and tsunami that followed resulted in the FDNPP accident that caused the release of a significant amount of radioactive materials from the reactors into the atmosphere. In this regard, the land near the FDNPP was highly contaminated, which heightened concern among residents living in proximity to the plant about the health and safety risks associated with the radioactive materials.

12.1.2 *Radiocesium Nuclides*

Among the radioactive materials deposited on the land, cesium-134 (^{134}Cs) and cesium-137 (^{137}Cs) were the primary sources of nuclides released after the accident [1]. The activity ratio of $^{134}Cs/^{137}Cs$ reported was 1:1 during the 2011 accident [2], though the half-lives of the nuclides are 2.1 years for ^{134}Cs and 30.2 years for ^{137}Cs. Therefore, both nuclides can affect the long-term external and internal radiation exposure of the inhabitants [1]. To overcome the problem of internal exposure, precautionary measures have been taken to address the elevated effects of radiocesium (^{134}Cs and ^{137}Cs) in agricultural products.

12.2 Regulation of Radiocesium in Food Monitoring System in Japan

12.2.1 *Provisional Regulation Values*

To maintain food safety in Japan, the government began to regulate the concentration of radionuclides in food on March 17, 2011. Prior to this, there was no regulation to monitor radionuclides for food grown in Japan. Thus, a provisional regulation for radioactive substances in food was adopted by the Japanese Ministry of Health, Labor, and Welfare (MHLW). The regulation set the annual maximum permissible dose of radiocesium in foods at 5 mSv, assuming the long-term consumption of food in Japan [3, 4]. In the regulation, limits for radiocesium, which is a combination of ^{137}Cs and ^{134}Cs, have been established, with the maximum permissible dose, 5 mSv/year, being split into 1 mSv/year for each of the 5 different food categories. The radiocesium limits were calculated by considering the food intake and its susceptibility to adults, young children, and infants (Table 12.1).

Table 12.1 Provisional regulation values for radiocesium and the new standard for radiocesium in foods in Japan (modified from [3])

Provisional regulation values[a]		New standard[b]	
Category	Limit (Bq/kg)	Category	Limit (Bq/kg)
Drinking water	200	Drinking water	10
Milk, dairy products	200	Milk	50
Vegetables	500	General foods	100
Grains	500	Infant foods	50
Meat, eggs, fish, etc.	500		

[a]Provisional regulation values were introduced on March 17, 2011
[b]The new standards were introduced on April 1, 2012

12.2.2 New Standards Limits

The Japanese government decided to reduce the maximum permissible dose from 5 mSv/year to 1 mSv/year on April 1, 2012. The categories were re-organized into 4 categories, including drinking water, milk, general foods, and infant foods. The limit for drinking water was established at 10 Bq/kg, consistent with the World Health Organization's guidance, which corresponded to approximately 0.1 mSv/year. The limit for general foods was calculated by taking the food intake and age into account, assuming that 50% of the market foods were contaminated [5]. As a result, the maximum limit among all age categories was 120 Bq/kg, rounded down to 100 Bq/kg. The new standard limits for milk and infant foods for children were established, and the MHLW set the new limit at 50 Bq/kg, assuming that all these foods were contaminated. Overall, the limit has been changed from 500 Bq/kg to 100 Bq/kg, corresponding to the change of permissible dose from 5 mSv/year to 1 mSv/year (Table 12.1).

12.2.3 Local Government Food Inspections

To ensure the safety of food in Japan, a food monitoring system was established (Fig. 12.1), and the prefectural governments were given responsibility for its implementation and day-to-day operation [6]. In short, the Nuclear Emergency Response Headquarters provides instructions, or the orders to the local governments in 17 prefectures (Aomori, Iwate, Akita, Miyagi, Yamagata, Fukushima, Ibaraki, Tochigi, Gunma, Chiba, Saitama, Tokyo, Kanagawa, Niigata, Yamanashi, Nagano, and Shizuoka). The local government performs all food inspections and reports the data to the MHLW. The local government also makes the inspection plans, collects samples, and measures radioactivity in the samples. As of 2018, in Fukushima Prefecture, there are currently 11 germanium semiconductor detectors in operation at the Fukushima Agricultural Technology Center.

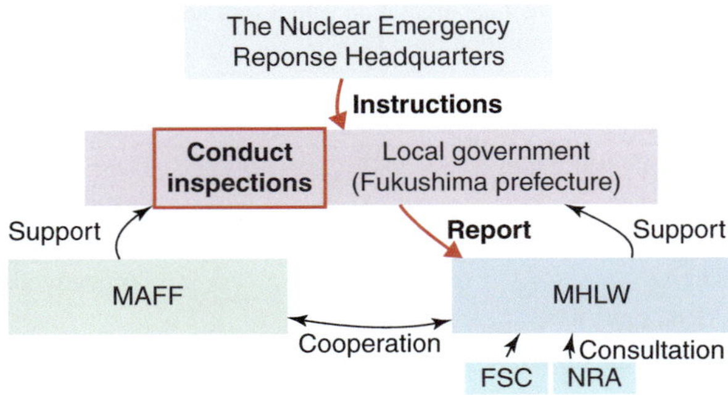

Fig. 12.1 Food monitoring system in Japan. Local government (Fukushima prefecture) is responsible for food inspections. *MAFF* Ministry of Agriculture, Forestry and Fisheries, *MHLW* Ministry of Health, Labor and Welfare, *FSC* Food Safety Commission, *NRA* Nuclear Regulation Authority

12.3 Time Course of Radiocesium Concentration in Agricultural Products in Fukushima

12.3.1 Food Monitoring in Fukushima Prefecture

Fukushima prefecture reports the monitoring data of between 1000 and 3000 samples per month because of the high anxiety associated with the close proximity to the accident site [7]. Figure 12.2 shows the time course of radiocesium concentration in foods produced in Fukushima in the first 5 years after the accident. The green line represents the world standard [9]; the red line represents the limit of the provisional regulation (from March 2011 to March 2012), and the limit of the new standard (from April 2012—present) (Table 12.1). In this study, radiation was not detected (N.D.) in most of the samples. For example, radiocesium contamination was not detected in 22,164 out of 23,604 (94%) vegetable specimens.

12.3.2 Cereals

Figure 12.2a shows the radiocesium concentration in cereals (excluding rice). From 2011 to 2016, 12,325 out of 12,469 (99%) specimens contained less than 100 Bq/kg. Analysis of the specimens showed that there was a relatively high level of contamination in the samples collected in 2011. The leaves and stems of winter crops were contaminated from the fallout in March 2011, and the harvested grain had more radiocesium when the surface area of foliage was large at the time of the accident [10]. Therefore, the cereals with high radiocesium concentrations in 2011 were contaminated directly from the fallout. In addition, soybeans harvested in 2011–2013 had a higher concentration of radiocesium [7, 11], suggesting indirect contamination from soil via the roots.

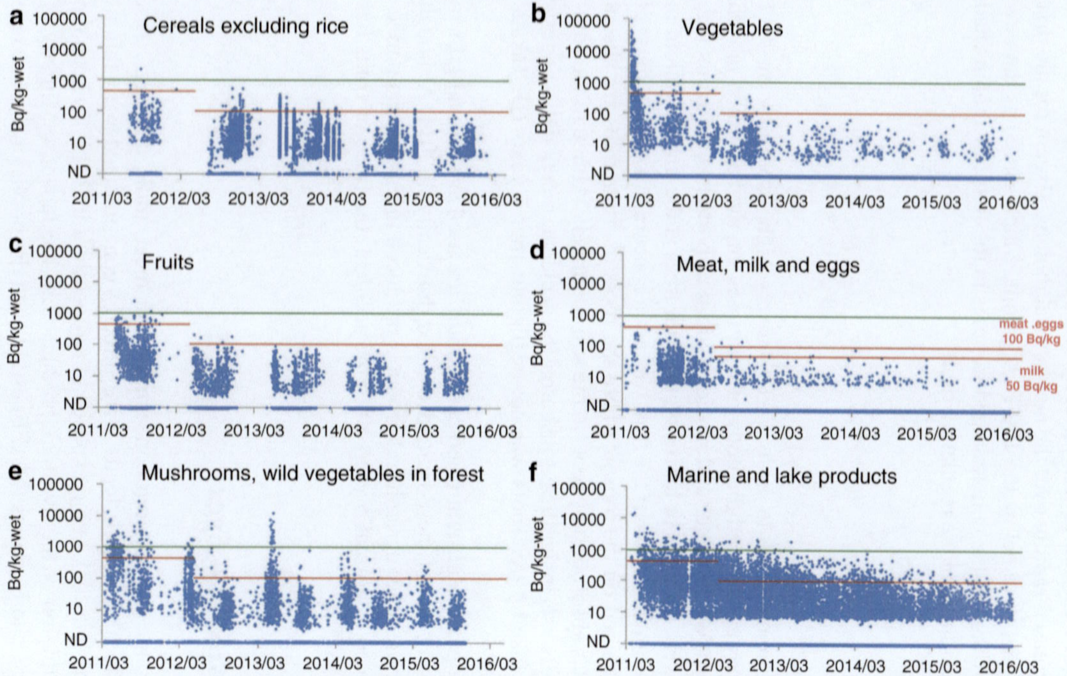

Fig. 12.2 Radiocesium concentrations in agricultural products in Fukushima prefecture from March 2011 to March 2016 (modified from [8]). In Japan, the concentration of radiocesium (^{137}Cs + ^{134}Cs) is employed and assigned to Bq/kg-wet in food monitoring. (**a**) Cereals excluding rice, (**b**) vegetables, (**c**) fruits, (**d**) meat, milk, and eggs, (**e**) mushrooms, wild vegetables in the forest, (**f**) marine and lake products. *ND* not detected. The green lines indicate the level of CODEX STAN 193-1995, and the red lines indicate the levels of regulation in Japan. Regulation of general food in Japan was changed from 500 Bq/kg to 100 Bq/kg in April 2012. In this study, we did not fit a trend-line because the radiation was not detected (N.D.) in most of the samples. The data shown in this figure is derived from only Fukushima prefecture, but other prefectures are also inspecting their agricultural products even though the scale is much smaller than that of Fukushima prefecture

12.3.3 Vegetables

Figure 12.2b shows the radiocesium concentrations in vegetables. Just after the accident, the vegetable specimens were seriously contaminated. Vegetables that grew in the field during the fallout were directly contaminated. However, the contamination level decreased rapidly over the next several months. Except for the first 2 years, all the samples were below the 100 Bq/kg limit, which served as the new Japanese standard. Although the contaminated samples in the first few months had much higher radiocesium than the regulation limit.

12.3.4 Fruits

The radiocesium contamination in fruits was relatively high, especially in 2011 (Fig. 12.2c). Radiocesium entered the body of fruit trees via the bark, then moved to xylem and phloem, where water and nutrients are transported to the fruits [12, 13]. Because there were no fruits or leaves on the fruit trees at the time of fallout in March 2011, direct contamination of the fruit was not a concern. Moreover, most of the radiocesium remained on the topsoil where the roots of fruits trees are absent [14], revealing that Cs absorption by roots in the soil is unlikely. Although the pathway from soil to fruits was insignificant just after the accident, recent data suggests that radiocesium uptake by tree roots was not negligible after 5 years (D. Takata, pers. comm.).

Fruits, especially peach and semi-dried persimmon, are used as seasonal gifts in Japan, and the market price of these fruits decreased drastically in 2011. Although the market price for the other agricultural products produced in Fukushima prefecture gradually returned to normal 5 years after the accident, the price of fruits has still not recovered [15].

12.3.5 Meat, Milk, and Eggs

Figure 12.2d shows the radiocesium concentration in meat, milk, and eggs. Just after the accident, local livestock feeds were directly contaminated with fallout which resulted in the livestock being exposed to radiocesium in 2011. To keep the radiocesium concentration of the livestock products low, the Japanese Ministry of Agriculture, Forestry, and Fisheries (MAFF) established the provisional allowance level of radiocesium for feeds on August 1, 2011. The radiocesium concentration was set at 300 Bq/kg for cow, cattle, pig, and poultry feeds, which was updated on

February 3, 2012, to 100 Bq/kg for cow, cattle, and pig feeds, and 80 Bq/kg for poultry feeds [16]. As a result of these strict allowance levels, radiocesium in livestock products (meat, milk, and eggs) was low over the last few years.

12.3.6 *Mushrooms and Wild Vegetables in the Forest*

The radiocesium concentration in mushrooms and wild forest vegetables has not decreased greatly (Fig. 12.2e). To compare this with the CODEX standard, 155 out of 6846 (2.2%) specimens exceeded 1000 Bq/g of radiocesium over the 5-year period (2011–2016). This result indicates that forests retain ^{137}Cs for a long time (Miura, unpublished data), and mushrooms are also known to retain ^{137}Cs over long periods of time [17]. Because it is impossible to remove ^{137}Cs from contaminated forests, it will take a long time to see a reduction in contamination. From an economic perspective, avoiding the distribution and consumption of wild forest vegetables is relatively easy. However, people who live in these mountainous areas are disappointed because eating seasonal wild plants is often part of their lifestyle.

Prior to the accident, Fukushima prefecture supplied most of the 25 million oak logs used annually for commercial shiitake (*Lentinus edodes*) log-cultivation in Japan [18]. Today, shiitake producers use only logs sourced from regions not affected by the radioactive fallout (T. Iizumi, pers. comm.). Nevertheless, there is a strict limit of radiocesium (^{134}Cs + ^{137}Cs, <50 Bq/kg) permitted in logs by the Japanese Forestry Agency [19] to ensure the concentration will be lower than the maximum tolerable level of radioactivity in general food (100 Bq/kg). Mushrooms are able to take up various stable or radioactive elements and accumulate them in their fruit bodies [20]. The ratio of an element in a fruit body to its concentration in a substrate such as wood, known as the transfer factor (TF), is a measure of the ability of a mushroom to accumulate an element. In March 2011, most of the radioactive fallout contaminated the tree bark and not the inner wood [21]. Since then, the ^{137}Cs concentration within trees has been in flux and therefore the TF of ^{137}Cs between logs and shiitake is not static. Cesium-137 ions will migrate from the bark into the wood and be taken up through the roots from the contaminated soil; ^{137}Cs will not become homogeneously distributed within trees for an undetermined number of years ($\geq$12 years based on data published by Yoshida et al. [22]). It is believed that shiitake mycelium primarily grows in the sapwood of logs [23], and an increase in radiocesium in the sapwood will result in a concomitant increase of radiocesium in shiitake mushrooms. In the coming years and decades, the distribution of radiocesium within oak trees felled from contaminated forests and radiocesium transfer to shiitake mushrooms during their cultivation using these logs will need to be closely monitored.

Table 12.2 Inspection of all rice produced in Fukushima prefecture

Production year	Number of rice bags[a]	Number of rice bags containing more than 100 Bq/kg of radiocesium (sum of ^{137}Cs and ^{134}Cs)
2012	10,346,169	71
2013	11,005,859	28
2014	11,014,941	2
2015	10,498,579	0
2016	10,265,957	0

[a]Rice bags contain 30 kg of brown rice (unpolished rice)

12.3.7 Marine and Lake Products

Figure 12.2f illustrates how radiocesium concentration in marine and lake products has decreased over time. The contamination level was found to be dependent on the types of sea and freshwater creatures analyzed [8]. In short, migratory fish and crustaceans were reported to contain low radiocesium concentrations. On the other hand, bottom-dwelling fish (such as flounder) and freshwater fish contained higher radiocesium concentrations, indicating that the seafloor and freshwater lakes had high concentrations of bioavailable radiocesium.

12.3.8 Rice

Rice is the main staple food in Japan, and it is strictly inspected to reduce the risk of internal exposure to radiocesium. Since 2012, a year after the accident, all the rice produced in Fukushima has been inspected by purpose-built monitoring equipment, known as belt conveyor testers that can inspect a 30 kg rice bag in less than 30 s. Nearly 200 belt conveyor testers are installed in Fukushima prefecture to inspect all rice after harvest in this prefecture [24]. As shown in Table 12.2, more than 10 million bags of rice are inspected each year with very few instances of contamination (Table 12.2). The inspections are still in progress as of 2019, although the government is considering to terminate these inspections in the near future (Fukushima prefecture, personal communication).

12.4 Countermeasures and Strategies for Reducing Radiocesium Concentration in Crops

12.4.1 Decontamination of Soil

MAFF [25] identified three primary countermeasures to reduce the risk of radiocesium contamination in crops: (1) removing the topsoil (stripping of topsoil), (2) deep ploughing to bury the topsoil (inversion tillage), and (3) removing the fine soil particle in paddy fields using water. When the radiocesium concentration in soil is more than 10,000 Bq/kg, removing the topsoil is recommended. When the radiocesium concentration ranges from 5,000 to 10,000 Bq/kg, the three countermeasures listed above can be used to reduce the radiocesium levels. When the radiocesium concentration is less than 5,000 Bq/kg, decontamination is not an obligation but an option.

12.4.2 Potassium (K) Fertilizer: The Relationship Between K in Soil and Cs in Crops

In the contaminated areas in Fukushima, rice farmers are obligated to apply K fertilizer to paddy fields [26]. When farmers grow other crops, K fertilizer is highly recommended [27]. It was demonstrated that high K concentration in soils contributes to not only reducing the Cs^+ uptake through roots, but also reducing the Cs^+ translocation from leaves to the rice grains [28].

12.4.3 Molecular Mechanism of Cs^+ Uptake and Translocation

Recently, the molecular mechanism of Cs^+ uptake through plant roots has been clarified. HAK5 is a protein located in the plasma membrane of Arabidopsis roots, and it is responsible for K uptake under low K conditions [29]. Moreover, *hak5* mutants of Arabidopsis have low concentrations of Cs in low K conditions, suggesting that HAK5 is responsible for Cs^+ uptake through Arabidopsis roots [29]. In rice, OsHAK1 was reported to perform the same function as HAK5 in Arabidopsis [30, 31]. In the *oshak1* mutant, Cs concentration was drastically reduced under low K condition, but unchanged under sufficient K condition [30]. The K concentration in *hak5* [29] and *oshak1* [30] decreased compared with the wild types, which caused biomass to decrease [30]. On the other hand, for the *oshak1* mutant that originated from another cultivar of rice (i.e., Akitakomachi), a decrease in both biomass and K concentration was not observed [31]. Therefore, OsHAK1 may become a breeding target to reduce the Cs concentration under K-deficient conditions. When K fertilizer is applied to soil, mutants of OsHAK1 are of little use because OsHAK1 occurs only under the K-deficient conditions.

There is no molecular information about the Cs translocation into grains, which could be a potentially important breeding target. It is necessary to reveal the mechanisms of Cs^+ translocation in plants. Aside from the Cs^+ transport, altering root architecture is also a breeding target. For example, a gene locus is reportedly responsible for deep roots in rice [32]. In the future, it may be essential to create new crops that have deeper roots, which can help to reduce Cs^+ absorption from the topsoil since Cs^+ remains in the topsoil and moves downward very slowly [33].

12.5 Conclusions

In Japan, the local government is responsible for food monitoring. From the inspection data collected from Fukushima prefecture, most of the samples are below the analytical detection limits, and very few samples exceed 100 Bq/kg, which is the Japanese standard for radiocesium. Potassium fertilizer application began shortly after the accident to reduce the ^{137}Cs concentrations in Fukushima crops. Efforts to produce low radiocesium crops continue. The gene candidates and molecular systems are currently in the development phase.

References

1. MEXT and MAFF. Overview of the results of survey research on the distribution of radioactive materials released by the Tokyo Electric Power Company Fukushima Daiichi Nuclear Power Plant accident (in Japanese text). 2012. http://radioactivity.nsr.go.jp/ja/contents/6000/5233/24/5233_20120313_20120615_rev20130701.pdf. Accessed 20 July 2018.
2. Merz S, Steinhauser G, Hamada N. Anthropogenic radionuclides in Japanese food: environmental and legal implications. Environ Sci Technol. 2013;47(3):1248–56.
3. Hosono H, Kumagai Y, Sekizaki T. Development of an information package of radiation risk in beef after the Fukushima Daiichi nuclear power plant accident. In: Nakanishi T, Tanoi K, editors. Agricultural implications of the Fukushima nuclear accident. Tokyo: Springer; 2013.
4. MHLW. Handling of food contaminated by radioactivity. Notice no. 0317 article 3 of the Department of Food Safety. 2011. https://www.mhlw.go.jp/stf/houdou/2r9852000001558e-img/2r98520000015av4.pdf. Accessed 20 July 2018.
5. MHLW. The ministerial ordinance partially revising the ministerial ordinance on milk and milk products concerning compositional standards, etc.; the notification on designating the radioactive substances designated by the Minister of Health, Labour and Welfare under the provisions of item (I) (1) of the attached table 2 of the ministerial ordinance on milk and milk products concerning compositional standards, etc.; and the Notification on partial revision of specification and standards for food, food additives, etc. Notice No. 0315 article 1 of the Department of Food Safety. 2012. https://www.mhlw.go.jp/english/topics/2011eq/dl/food-120821_1.pdf. Accessed 20 July 2018.
6. MHLW. Radioactive materials in foods. Current situation and protective measures. 2013. https://www.mhlw.go.jp/english/topics/2011eq/dl/food-130926_1.pdf. Accessed 20 July 2018.
7. Nihei N. Monitoring inspection for radioactive substances in agricultural, livestock, forest and fishery products in Fukushima prefecture. In: Nakanishi T, Tanoi K, editors. Agricultural implications of the Fukushima nuclear accident: the first three years. Tokyo: Springer; 2016.
8. Nihei N, Tanoi K, Nakanishi TM. Monitoring inspection for radiocesium in agricultural, livestock, forestry and fishery products in Fukushima prefecture. J Radioanal Nucl Chem. 2015;307(3):2217–20.
9. CODEX STAN 193-1995. General standard for contaminants and toxins in food and feed. 1995. http://files.foodmate.com/down.php?auth=AGEGYwhlAWQ. Accessed 20 July 2018.
10. Tanoi K. Behavior of radiocesium adsorbed by the leaves and stems of wheat plant during the first year after the Fukushima Daiichi nuclear power plant accident. In: Nakanishi T, Tanoi K, editors. Agricultural implications of the Fukushima nuclear accident. Tokyo: Springer; 2013.
11. Nihei N. Radioactivity in agricultural products in Fukushima. In: Nakanishi T, Tanoi K, editors. Agricultural implications of the Fukushima nuclear accident. Tokyo: Springer; 2013.

12. Takata D, Yasunaga E, Tanoi K, Nakanishi TM, Sasaki H, Oshita S. Radioactivity distribution of the fruit trees ascribable to radioactive fall out (IV): caesium content and its distribution in peach trees. Radioisotopes. 2012;61(12):607–12.
13. Takata D. Distribution of radiocesium from the radioactive fallout in fruit trees. In: Nakanishi T, Tanoi K, editors. Agricultural implications of the Fukushima nuclear accident. Tokyo: Springer; 2013.
14. Sato M, Takata D, Tanoi K, Ohtsuki T, Muramatsu Y. Radiocesium transfer into the fruit of deciduous fruit trees contaminated during dormancy. Soil Sci Plant Nutr. 2015;61(1):156–64.
15. MAFF. Report of distribution of agricultural products from Fukushima prefecture in 2017 (in Japanese text). 2018. http://www.maff.go.jp/j/press/shokusan/ryutu/attach/pdf/180328-1.pdf. Accessed 20 July 2018.
16. MAFF. Allowance level of radiocesium in fertilizer, soil improvement materials, soil supplement, and feed (in Japanese text). 2012. http://www.maff.go.jp/j/syouan/soumu/saigai/pdf/110801_tsuchi_tokekomi.pdf. Accessed 20 July 2018.
17. Muramatsu Y, Yoshida S, Sumiya M. Concentrations of radiocesium and potassium in basidiomycetes collected in Japan. Sci Total Environ. 1991;105:29–39.
18. Miura S. The effects of radioactive contamination on the forestry industry and commercial mushroom-log production in Fukushima, Japan. In: Agricultural implications of the Fukushima nuclear accident. Tokyo: Springer; 2016. p. 145–60.
19. Forestry Agency. Provisional regulation value of radiocesium in mushroom log and mushroom bed (in Japanese text). 2012. http://www.rinya.maff.go.jp/j/tokuyou/shiitake/pdf/120830_1.pdf. Accessed 20 July 2018.
20. Falandysz J, Borovička J. Macro and trace mineral constituents and radionuclides in mushrooms: health benefits and risks. Appl Microbiol Biotechnol. 2013;97(2):477–501.
21. Kuroda K, Kagawa A, Tonosaki M. Radiocesium concentrations in the bark, sapwood and heartwood of three tree species collected at Fukushima forests half a year after the Fukushima Dai-ichi nuclear accident. J Environ Radioact. 2013;122:37–42.
22. Yoshida S, Watanabe M, Suzuki A. Distribution of radiocesium and stable elements within a pine tree. Radiat Prot Dosim. 2011;146(1-3):326–9.
23. Tokimoto K. Shiitake log cultivation. In: Shiitake cultivation - mushroom growers' handbook 2. Seoul: MushWorld; 2005, chap. 3, pp. 46–60.
24. Nihei N, Tanoi K, Nakanishi TM. Inspections of radiocesium concentration levels in rice from Fukushima prefecture after the Fukushima Dai-ichi nuclear power plant accident. Sci Rep. 2015;5(1):8653.
25. MAFF. Technical report of agricultural land decontamination countermeasures (in Japanese text). 2013. http://www.maff.go.jp/j/nousin/seko/josen/pdf/gaiyou.pdf. Accessed 20 July 2018.
26. Fukushima Prefecture. Potassium application on paddy field to reduce radiocesium in rice (in Japanese text). 2012. https://www.pref.fukushima.lg.jp/download/1/ganba24H240410.pdf. Accessed 20 July 2018.
27. Fukushima Prefecture. Potassium application on agricultural field to reduce radiocesium in soybean and buckwheat (in Japanese text). 2013. https://www.pref.fukushima.lg.jp/download/1/future-38daizusobataisakuH250325.pdf. Accessed 20 July 2018.
28. Nobori T, Kobayashi NI, Tanoi K, Nakanishi TM. Effects of potassium in reducing the radiocesium translocation to grain in rice. Soil Sci Plant Nutr. 2014;60(6):772–81.
29. Qi Z, Hampton CR, Shin R, Barkla BJ, White PJ, Schachtman DP. The high affinity K+ transporter AtHAK5 plays a physiological role in planta at very low K+ concentrations and provides a caesium uptake pathway in Arabidopsis. J Exp Bot. 2008;59(3):595–607.
30. Nieves-Cordones M, Mohamed S, Tanoi K, Kobayashi NI, Takagi K, Vernet A, Guiderdoni E, Périn C, Sentenac H, Véry A-A. Production of low-Cs^+ rice plants by inactivation of the K^+ transporter OsHAK1 with the CRISPR-Cas system. Plant J. 2017;92(1):43–56.
31. Rai H, Yokoyama S, Satoh-Nagasawa N, Furukawa J, Nomi T, Ito Y, Fujimura S, Takahashi H, Suzuki R, Yousra E, Goto A, Fuji S, Nakamura S-I, Shinano T, Nagasawa N, Wabiko H, Hattori

H. Caesium uptake by rice roots largely depends upon a single gene, HAK1, which encodes a potassium transporter. Plant Cell Physiol. 2017;58(9):1486–93.
32. Uga Y, Okuno K, Yano M. Dro1, a major QTL involved in deep rooting of rice under upland field conditions. J Exp Bot. 2011;62(8):2485–94.
33. Wakiyama Y, Onda Y, Mizugaki S, Asai H, Hiramatsu S. Soil erosion rates on forested mountain hillslopes estimated using ^{137}Cs and $^{210}Pb_{ex}$. Geoderma. 2010;159(1-2):39–52.

Chapter 13
Isotopic Signatures of Actinides in Environmental Samples Contaminated by the Fukushima Daiichi Nuclear Power Plant Accident

Aya Sakaguchi and Georg Steinhauser

Abstract Large amounts of radionuclides were released into the environment as a result of the Fukushima Daiichi Nuclear Power Plant (FDNPP) accident following the magnitude 9.0 Great East Japan Earthquake and subsequent tsunami on March 11, 2011. Some research activities that focused on the monitoring of actinide elements (^{236}U, ^{238,239,240,241}Pu, ^{241}Am, and ^{242,243,244}Cm) in road dust (black substances), litter, and soil samples from Fukushima prefecture are summarized in this chapter.

Keywords Radioactivity · Isotope · Actinides · Volatile element · Refractory element · Environmental samples · Black substance

13.1 Introduction

The magnitude 9.0 Great East Japan Earthquake occurred in the northwest Pacific on 11th March 2011; the earthquake was followed by a massive tsunami, the damage from which led to the severe accident at the Fukushima Daiichi Nuclear Power Plant (FDNPP) and the release of large amounts of radionuclides into the environment. From broad survey work and subsequent investigations on the emitted radionuclides via radiological measurement of environmental samples, analyses of biological samples, and atmospheric dispersion simulations, a clear picture about the level and spread of contaminant radionuclides, especially iodine-131 (^{131}I), cesium-134 and -137 (^{134}Cs and ^{137}Cs), and exposure doses to communities started

A. Sakaguchi (✉)
Center for Research in Isotopes and Environmental Dynamics, University of Tsukuba, Tsukuba, Ibaraki, Japan
e-mail: ayaskgc@ied.tsukuba.ac.jp

G. Steinhauser
Leibniz Universität Hannover, Institute of Radioecology and Radiation Protection, Hannover, Germany

G. Steinhauser et al. (eds.), *Nuclear Emergencies*,
Current Topics in Environmental Health and Preventive Medicine,
https://doi.org/10.1007/978-981-13-8327-4_13

to emerge. However, potential contamination by release of uranium (U) and transuranic nuclides such as plutonium, americium, and curium (Pu, Am, and Cm) from the fuel cores also started to cause concern due in part to their measurement in a limited number of samples and also because of the difficulties associated with trace analysis of such α-emitting radionuclides. Due to the use of MOX fuel which contains about 6% by weight of Pu in plant No. 3, in addition to the production of Pu isotopes from spent U fuel, residents in the immediate neighbourhood of the FDNPP became extremely apprehensive about the possible release of Pu from the plant (cf. interviews of citizens from Okuma town).

Although it might be thought that discharge of actinides, especially the transuranic elements, is uncommon, these elements have been continuously released into the environment, mainly during the 1950s to 1960s (global fallout), due to atmospheric nuclear testing leakages/emissions from nuclear facilities, and nuclear accidents. Most of these actinides are produced by the sequential/multiple neutron captures of $^{235,238}U$ and ^{239}Pu during the burning (neutron irradiation) of fuels in nuclear power stations. The chemically toxic actinides emit biologically hazardous α-particles/rays, which have a large radiation weighting factor 20 (cf. 1 for X-rays and γ-rays). In addition, they have long half-lives and they exist in a variety of chemical oxidation states (i.e., valences +III – +VI) in the environment. Thus, it is essential to know the concentrations, distributions, and behaviour of these elements in the environment and to understand their effects on humans, animals, and plant systems.

The purpose of recent research has been to obtain accurate elemental and isotopic information on the FDNPP-derived uranium and transuranic nuclides in the environment. In this context, U isotopes (^{236}U, ^{238}U), Pu isotopes (^{238}Pu, ^{239}Pu, and ^{240}Pu), Am and Cm isotopes (^{241}Am, ^{242}Cm, ^{243}Cm, and ^{244}Cm) were measured in road dust samples (black substances), litter, and soil samples; many of the samples were collected at the most heavily contaminated areas in Fukushima Prefecture. The magnitude of the emissions together with the isotopic compositional data is discussed in relation to the fuel core inventories of the FDNPP obtained from the ORIGEN 2 model simulation by Nishihara et al. [1].

13.2 Isotopes of Interest

13.2.1 Uranium

Three isotopes of uranium occur naturally on Earth in macroscopic amounts, namely ^{238}U (99.2742%; $T_{1/2} = 4.468 \cdot 10^9$ years), fissile ^{235}U (0.7204%; $T_{1/2} = 7.038 \cdot 10^8$ years), and ^{234}U (0.0054%; $T_{1/2} = 2.455 \cdot 10^5$ years), the latter of which is a decay product of ^{238}U (member of the ^{238}U decay chain). Uranium is not a particularly rare element in the upper continental crust (2.8 mg kg^{-1}) [2]. Consequently, a release of minute amounts of anthropogenically enriched uranium is difficult to discern by a shift in the uranium-isotopic characteristics of an environmental sample, given the relatively high levels of ubiquitous natural uranium.

Instead, ^{236}U ($T_{1/2} = 2.342 \cdot 10^7$ years) is a suitable and reliable indicator of anthropogenic uranium as it is primarily produced in a nuclear reactor upon neutron irradiation of uranium fuel, mainly via the nuclear reaction $^{235}U(n,\gamma)^{236}U$.

13.2.2 Neptunium

Neutron capture of ^{236}U in an (n,γ)-type reaction yields short-lived ^{237}U, which decays to ^{237}Np ($T_{1/2} = 2.144 \cdot 10^6$ years). An alternative route of production is the nuclear reaction $^{238}U(n,2n)^{237}U$, again, followed by beta decay to ^{237}Np. Although it constitutes a major constituent of spent nuclear fuel, no release of ^{237}Np was confirmed after the Fukushima nuclear accident, despite methods that have been developed for that purpose [3]. Early indications of a release of short-lived ^{239}Np [4] probably were a misinterpretation due to spectral interference with ^{129m}Te in gamma spectrometry [5]. Until the complete decommissioning of the FDNPP plant, the leaching of radionuclides, in particular actinides (including neptunium) from damaged and molten spent fuel in seawater must be watched closely [6].

13.2.3 Plutonium

Most plutonium isotopes, in particular ^{239}Pu ($T_{1/2}$ = 24 110 years), ^{240}Pu ($T_{1/2}$ = 6563 years), ^{241}Pu ($T_{1/2}$ = 14.35 years), and ^{242}Pu ($T_{1/2} = 3.750 \cdot 10^5$ years), are produced by neutron capture in ^{238}U, yielding short-lived ^{239}U, which decays to ^{239}Np and subsequently to ^{239}Pu. Further neutron captures yield heavier isotopes of plutonium. The onset of ^{238}Pu ($T_{1/2}$ = 87.74 years) is mostly due to neutron capture in ^{237}Np, followed by beta decay of the ^{238}Np. The most prominent isotopes are certainly (fissile) ^{239}Pu and its activation product ^{240}Pu. Both nuclides are α-emitters, but since their α-energies are almost identical, both nuclides cannot be differentiated in regular alpha spectrometry. Literature values, therefore, are often given as sum activities of both $^{239+240}Pu$. The isotopic ratio of $^{240}Pu/^{239}Pu$ is the most reliable indicator for the distinction between nuclear weapons fallout-related plutonium and reactor plutonium. Global fallout plutonium is characterized by an isotopic ratio of $^{240}Pu/^{239}Pu$ of 0.176 ± 0.014 [7]. However, this global, integral range may be shifted locally. For example, the Pu signature from the Pacific Proving Grounds is significantly higher due to high yield explosions that can still be found in the environment [8, 9], up to 0.363 ± 0.004 [10]. Since alpha spectrometry cannot be applied for the distinction of the isotopic ratio, mass spectrometric methods have to be applied. A great variety of methods exists, from straightforward inductively coupled plasma mass spectrometry (ICP-MS) to more sophisticated accelerator mass spectrometry (AMS), which represents the gold standard of techniques available with respect to

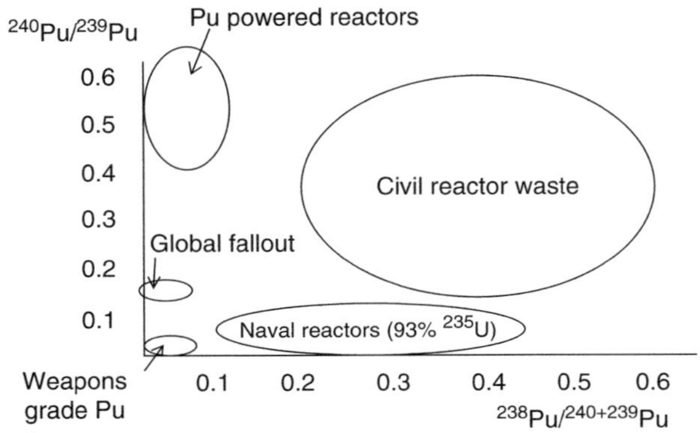

Fig. 13.1 Isotopic signatures of plutonium for various relevant sources. Taken from Cagno et al. [11] and reprinted with permission from the Royal Society of Chemistry © 2014

sensitivity and suppression of isobaric interferences. If radiometric methods have to be used, the ratio of ^{238}Pu/$^{240+239}$Pu is used, which can be measured by alpha spectrometry. Figure 13.1 shows the various sources and their respective "plutonium fingerprint" [11].

13.2.4 Americium and Curium

Relatively short-lived ^{241}Pu decays to ^{241}Am ($T_{1/2}$ = 432.2 years). Americium-241 is a powerful α-emitter, but is also one of the few potent γ-emitters amongst environmental transuranic elements. Neutron capture in ^{241}Am yields short-lived ^{242}Am, which then decays to ^{242}Cm ($T_{1/2}$ = 162.94 days), which then decays to ^{238}Pu. Most relevant curium isotopes are shorter-lived than many other environmental actinides with $T_{1/2}$ = 29.1 years for ^{243}Cm and $T_{1/2}$ = 18.10 years for ^{244}Cm. Americium and all further (heavier) actinides (with the exception of nobelium) are primarily trivalent and thus potentially more immobile than ions with higher valence such as +V and +VI that typically form oxo-cations. At the same time, trivalent actinides can be more mobile and bioavailable than, for example, tetravalent plutonium.

13.3 Measurements of Actinides

Japanese scholars have a long history in analysing U, Pu, Am, and Cm isotopes sequentially, e.g., by using the methods of Sakaguchi et al. [12–14] and Yamamoto et al. [15–19]. The samples are heated to 450 °C and thereby calcinated. Note that the calcination of soil samples at higher temperatures may trap the plutonium fraction in newly built mineral phases [20]. The analytes are then leached from the ashed residue using solutions of mixed acids. For the quantification of the measured amount of the analyte, a known amount of ^{242}Pu and ^{243}Am is added to sample, because background levels of these nuclides are very low. For the quantification of ^{236}U, the $^{236}U/^{238}U$ ratio is used and natural ^{238}U serves as an internal standard. Purification of the radionuclide solution is usually performed using extraction chromatographic or ion exchange methods. After further chemical separation, α-spectroscopy is usually performed using surface barrier Si detectors.

In the case of selected samples of the black substances and litter samples, the U and Pu fractions were freshly separated and purified without adding any tracers [19]. General information on mass spectrometry for radionuclide applications can be found elsewhere [21].

13.4 Detections of Actinides

The first detection of plutonium releases from Fukushima Daiichi (with isotopic evidence) was achieved by Zheng et al. [22] using sector field mass spectrometry. Two litter and one surface soil sample contained plutonium that carried a distinct reactor signature ($^{240}Pu/^{239}Pu$ isotopic ratio >0.3, and exceptionally high relative content of ^{241}Pu, which may become long-term dose relevant). The authors concluded that $1.0 \cdot 10^9$–$2.4 \cdot 10^9$ Bq $^{239+240}Pu$ and $1.1 \cdot 10^{11}$–$2.6 \cdot 10^{11}$ ^{241}Pu have been released in the course of the accident. A second evidence provided by Schneider et al. [23] further proved not only the release of Fukushima-derived plutonium, but also the fact that the distinction of Fukushima-plutonium from global fallout plutonium is notoriously difficult and preferably requires samples that exhibit a low background in Pu. In that study, it was mainly vegetation samples from 2011 that allowed for the identification of the reactor plutonium signature.

Given the above-described challenges, it appeared prudent to use soil samples mainly for evaluating inventories (i.e., the accumulated levels) of the Fukushima-derived transuranic nuclides, while analysis of black substances and litter samples was used to gain information on accurate isotopic compositions for the uranium and transuranic nuclides released into the environment [19].

13.4.1 Cs and Pu Inventories in Soil, and the Fukushima-Derived Pu

Our previous study [19] revealed exceptionally high radiocesium contaminations/depositions in most samples. The distinct $^{134}Cs/^{137}Cs$ activity ratio of just below 1 [24] evidenced that virtually the entity of the radiocesium stemmed from the Fukushima nuclear accident. In contrast, depositions of $^{239,240}Pu$ were relatively low, yet detectable, ranging from 0.24 to 55 Bq/m^2. The fingerprint of $^{238}Pu/^{239,240}Pu$ activity ratios was 0.024–1.19 [19]. Those values with a $^{238}Pu/^{239,240}Pu$ activity ratio >0.03 were apparently affected by the Fukushima Daiichi accident and transuranic releases from one of its reactor units (see Fig. 13.1).

Our in-depth analyses [19] showed that the FDNPP released a relatively small amount of plutonium isotopes over a wide area. The $^{239,240}Pu/^{137}Cs$ activity ratios covered a wide range from 2.6×10^{-9} to 4.8×10^{-6} with mean and median values of $(2.1 \pm 6.4) \times 10^{-7}$ and 5.7×10^{-8}, respectively, see Table 13.1.

13.4.2 Isotopic Uranium, Plutonium, Americium, and Curium Signatures for the Black Substances and Litter Samples

More than 100 samples from various areas within the 20 km evacuation zone were analysed for ^{236}U, ^{238}Pu, $^{239,240}Pu$, ^{241}Am, ^{242}Cm, and $^{243,244}Cm$ by AMS, ICP-MS, and α-spectrometry after radiochemical separation [19]. However, a detection of Am and Cm was not possible in the soil samples. The activity ratios are summarized in Table 13.1 and compared with the core inventories estimated using the ORIGEN2 code [1].

13.4.2.1 Plutonium Isotopes

Like several previous studies, we showed a rather large variation range of $^{239,240}Pu$ concentrations in the black substances of 0.013–3.92 Bq/kg ($n = 105$) with arithmetic mean and median values of 0.33 ± 0.48 and 0.20 Bq/kg, respectively (see Table 13.1). The $^{238}Pu/^{239,240}Pu$ activity ratios allowed for a distinction from global fallout as they were higher for most samples compared to the nuclear weapons fallout. Activity ratios >2, however, were not observed in soil samples, though.

In contrast to the results presented by Schneider et al. [23], there was no apparent dependence of the ratios on distance or direction from the plant.

The isotopic $^{240}Pu/^{239}Pu$ ratios confirmed a rather narrow range of 0.31–0.37 ($n = 9$) (Table 13.1). Since these ratios were higher than the generally agreed-up ratio of global nuclear weapons fallout (0.18), this finding confirmed the origin of the contamination as reactor plutonium. Further discussion on this topic is given in Chap. 14.

13.4.2.2 Americium

Given its formation history of beta decay of ^{241}Pu, most of the fallout-derived ^{241}Am would attain an activity ratio of $^{241}Am/^{239,240}Pu$ in the soil with a value of about 0.4 (as of March 2011) [15]. As for the black substances, ^{241}Am concentrations were detected in the range 0.01–2.44 Bq/kg ($n = 94$). The $^{241}Am/^{239,240}Pu$ activity ratios for all samples were in the range 0.27–1.30 with mean and median values of 0.56 ± 0.16 and 0.54, respectively. This mean value is approximately two times higher than the core inventory value [1]; further work is needed to clarify this discrepancy.

Table 13.1 Comparison of activity ratios (or atomic ratios) amongst transuranic nuclides between core inventories of the Fukushima Daiichi reactor units 1, 2, and 3, and environmental samples (black substances and litter samples) [19]

	Inventories in the fuel core[a]			Activity ratios in environmental samples			Activity ratios in environmental samples[b]		
Activity ratio	Unit 1	Unit 2	Unit 3	*n*	Average	Median	*n*	Average	Median
$^{239+240}Pu/^{137}Cs$	$7.86 \cdot 10^{-3}$	$7.54 \cdot 10^{-3}$	$9.96 \cdot 10^{-3}$	137	$(1.20 \pm 1.31) \cdot 10^{-7}$	$0.79 \cdot 10^{-7}$	137	$(1.20 \pm 1.31) \cdot 10^{-7}$	$0.79 \cdot 10^{-7}$
$^{236}U/^{239+240}Pu$	$3.63 \cdot 10^{-4}$	$3.66 \cdot 10^{-4}$	$2.84 \cdot 10^{-4}$	12	$(5.84 \pm 4.43) \cdot 10^{-4}$	$5.12 \cdot 10^{-4}$	12	$(5.47 \pm 4.55) \cdot 10^{-4}$	$4.43 \cdot 10^{-4}$
$^{238}Pu/^{239+240}Pu$	2.90	2.38	2.30	137	1.73 ± 0.75	1.79		2.53	
$^{241}Am/^{239+240}Pu$	0.35	0.23	0.23	122	0.56 ± 0.16	0.54	96	0.46 ± 0.34	0.39
$^{242}Cm^{c}/^{239+240}Pu$	56.1	46.5	43.3	122	28.1 ± 13.1	26.7	97	40.5 ± 21.4	37.7
$^{242}Cm^{c}/^{234+244}Cm$	32.5	27.5	20.0	122	28.3 ± 9.9	25	122	28.3 ± 9.9	25
$^{243+244}Cm/^{239+240}Pu$	1.72	1.68	2.16	122	1.10 ± 0.61	1.06	97	1.54 ± 0.87	1.47
Atomic ratio									
$^{240}Pu/^{239}Pu$	0.34	0.32	0.36	9	0.33 ± 0.02	0.34	9	0.33 ± 0.02	0.34

Reprinted with permission from Elsevier © 2014

[a]Taken from Nishihara et al. [1]

[b]Calculated values using $^{238}Pu/^{239+240}Pu = 2.53$

[c]Decay-corrected to the date of the accident on March 11, 2011

It is of interest to discuss the growth and decay of ^{241}Am following the decay scheme of $^{241}Pu \rightarrow {}^{241}Am \rightarrow {}^{237}Np$. Considering the global fallout $^{241}Pu/^{239,240}Pu$ activity ratio in the 1960s was 14 [15] and the activity ratios of $^{241}Pu/^{239,240}Pu$ and $^{241}Am/^{239,240}Pu$ for the core inventories were 100 and 0.54 [19, 22], respectively, on the date of the FDNPP accident, the contribution of ^{241}Am global fallout represents largest contribution for ^{241}Am in the environment. Subsequently, the growth of ^{241}Am from ^{241}Pu derived from the FDNPP accident, and ^{241}Am from the fuel core contributes to environmental ^{241}Am load [25]. The global fallout for the $^{241}Am/^{239,240}Pu$ activity ratio will achieve a maximum value of ca. 0.41 in 2033. Even though about 1% of Pu was newly added from the FDNPP accident to the Pu global fallout level, this ratio will be about 0.44 in 2045 and this additional contribution is not substantial. The $^{241}Am/^{239,240}Pu$ activity ratio which is composed of the Am and Pu derived from the FDNPP accident will become ca. 3.4 in 2080.

13.4.2.3 Curium Isotopes

After decay correction to March 2011, ^{242}Cm concentrations in black substances ranged from 0.21 to 57.0 Bq/kg ($n = 105$). The ^{242}Cm activity was the highest of all transuranic nuclides that have been analysed. The $^{243,244}Cm$ concentrations in black substances covered a wide range from 0.005 to 2.78 Bq/kg ($n = 94$), while in litter samples, they were in the range from 0.02 to 15.15 Bq/kg-ash ($n = 28$).

The $^{242}Cm/^{243,244}Cm$ activity ratios in black substances ranged from 8.8 to 59.4 ($n = 94$) [19]. It is possible to clearly observe a variation in the $^{242}Cm/^{243,244}Cm$ activity ratios (as of March 11, 2011) for the samples when the results are arranged in ascending order (Fig. 13.2a). Approximately a half of the values are found within those of between reactor core 2 (27.6) and core 3 (20.0) ($n = 45$), then they show a dramatic increase. The $^{242}Cm/^{243,244}Cm$ activity ratios for the southern area showed larger values overall, with large scattering when results were re-formatted according to the sampling location (Fig. 13.2b). In fact, the origin of radiocesium for the southern area, including Tokyo, is thought to be reactor 2. However, due to the disparity in the ratios of Cm in reactor cores 2 and 3 and reactor core 1 ($^{242}Cm/^{243,244}Cm = 32.5$), we were not able to conclude that the Cm (and probably other actinides) in the southern area was derived from reactor 2. Further research is needed here.

A potentially important feature concerning Cm isotopes is the growth of ^{238}Pu from ^{242}Cm. However, the contribution of ^{238}Pu from ^{242}Cm was negligible. In addition, the ^{242}Cm produced from the decay of ^{242m}Am was also negligibly small as discussed by [19].

13.4.2.4 Uranium Isotopes

Measurement of the $^{235}U/^{238}U$ ratio has provided evidence for anthropogenic U contamination of the environment. However, due to the high prevalence of naturally occurring ^{238}U (~a few ppm) in environmental samples, the detection of the

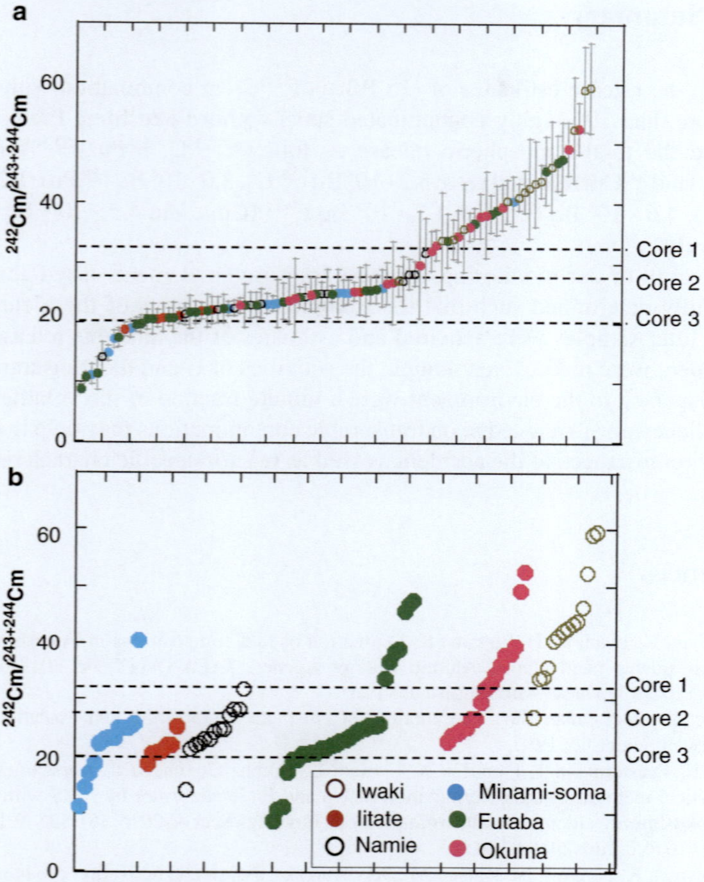

Fig. 13.2 ^{242}Cm/$^{243+244}$Cm activity ratios observed in samples obtained from each area and comparison with core inventories of Cm isotopes [1]

anthropogenic U is challenging. Given this situation, we have focused on the measurement of ^{236}U that is produced by reactions of ^{235}U(n,γ)^{236}U and ^{238}U(n,3n)^{236}U in nuclear reactors. To avoid the interference from global fallout of ^{236}U [12], some black substances ($n = 12$) which exhibited relatively high ^{238}Pu/^{239,240}Pu ratios were selected for analysis. As a result, very low concentrations of ^{236}U derived from the FDNPP, 0.28–6.74 × 10^{-4} Bq/kg, were successfully measured. The ^{236}U/^{239,240}Pu values in the black substances, 1.96–18.4 × 10^{-4} (weighted average 7.87 × 10^{-4}), were about 7 times higher than that for global fallout. These values were consistent with that found for the core inventory 2.84–3.63 × 10^{-4} [1], and thus provides evidence that fuel U and Pu fine particles were dispersed in the environment without undergoing large fractionation. An estimate of the dispersed U and Pu was calculated as 6 × 10^{-5}% of the total core inventory.

13.5 Summary

Based on the release estimates of (15 PBq) of ^{137}Cs in combination with samples from more than 100 highly contaminated spots within Fukushima Prefecture, we estimated the total atmospheric release as follows: ^{236}U, ^{238}Pu, $^{239,240}Pu$, ^{241}Am, $^{243,244}Cm$, and ^{242}Cm released were $5.2 \cdot 10^5$ Bq (^{236}U), $3.0 \cdot 10^9$ Bq (^{238}Pu), $1.2 \cdot 10^9$ Bq ($^{239,240}Pu$), $4.6 \cdot 10^8$ Bq (^{241}Am), $1.7 \cdot 10^9$ Bq ($^{243,244}Cm$), and $4.5 \cdot 10^{10}$ Bq (^{242}Cm), respectively.

As a result of our monitoring, all these transuranic elements listed above were successfully determined such that the isotopic compositions of these elements in soil and litter samples were acquired and estimates of the amounts released to the environment were made. Even though, the activities of U and the transuranium elements dispersed to the environment were a minute fraction of the volatile releases (e.g., radiocesium), knowledge on transuranic contaminations may help in elucidating the circumstances of the accident as well as reactor-specific characteristics.

References

1. Nishihara K, Iwamoto H, Suyama K. Estimation of fuel compositions in Fukushima Daiichi nuclear power plant. Japan Atomic Energy Agency. JAEA-Data/Code 2012-018. 2012, pp. 1–190 (in Japanese with English abstract).
2. Taylor SR, McLennan SM. The continental crust: its composition and evolution. Oxford: Blackwell Scientific; 1985.
3. Hain K, Faestermann T, Famulok N, Fimiani L, Gómez-Guzmán JM, Korschinek G, et al. Analytical method for the determination of Np and Pu in sea water by AMS with respect to the Fukushima accident. Nucl Instrum Methods Phys Res Sect A. 2015;361:505–9. https://doi.org/10.1016/j.nimb.2015.04.018.
4. Shozugawa K, Nogawa N, Matsuo M. Deposition of fission and activation products after the Fukushima Daiichi nuclear power plant accident. Environ Pollut. 2012;163:243–7. https://doi.org/10.1016/j.envpol.2012.01.001.
5. Rosenberg BL, Shozugawa K, Steinhauser G. Detection of fuel release in a nuclear accident: a method for preconcentration and isolation of reactor-borne ^{239}Np using ion-specific extraction chromatography. Anal Chem. 2015;87:8651–6. https://doi.org/10.1021/acs.analchem.5b02265.
6. Kirishima A, Hirano M, Akiyama D, Sasaki T, Sato N. Study on the leaching behavior of actinides from nuclear fuel debris. J Nucl Mater. 2018;502:169–76. https://doi.org/10.1016/j.jnucmat.2018.02.018.
7. Krey PW, Hardy EP, Pachucke H, Rourke F, Couluzza J, Benson WK. Mass isotopic composition of global fallout Pu in soil. In: Transuranium nuclides in the environment. STI/PUB/410. Vienna: IAEA; 1976. p. 672–8.
8. Pittauer D, Tims SG, Froehlich MB, Fifield LK, Wallner A, McNeil SD, et al. Continuous transport of Pacific-derived anthropogenic radionuclides towards the Indian Ocean. Sci Rep. 2017;7:44679. https://doi.org/10.1038/srep44679.
9. Wu J, Dai M, Xu Y, Zheng J. Plutonium in the western North Pacific: transport along the Kuroshio and implication for the impact of Fukushima Daiichi Nuclear Power Plant accident. Chem Geol. 2019;511:256. https://doi.org/10.1016/j.chemgeo.2018.12.006.
10. Pyle GL, Ghiorso A, Thompson SG, Higgins GH, Seaborg GT, Browne CI, et al. Heavy isotope abundances in mike thermonuclear device. Phys Rev. 1960;119:2000–4. https://doi.org/10.1103/PhysRev.119.2000.

11. Cagno S, Hellemans K, Lind OC, Skipperud L, Janssens K, Salbu B. LA-ICP-MS for Pu source identification at Mayak PA, the Urals, Russia. Environ Sci Processes Impacts. 2014;16:306–12. https://doi.org/10.1039/C3EM00531C.
12. Sakaguchi A, Kawai K, Steier P, Quinto F, Mino K, Tomita J, Hoshi M, Whitehead N, Yamamoto M. First results on ^{236}U levels in global fallout. Sci Total Environ. 2009;407:4238–42. https://doi.org/10.1016/j.scitotenv.2009.01.058.
13. Sakaguchi A, Kawai K, Steier P, Imanaka T, Hoshi M, Endo S, Zhumadilov K, Yamamoto M. Feasibility of using ^{236}U to reconstruct close-in fallout deposition from the Hiroshima atomic bomb. Sci Total Environ. 2010;408:5392–8. https://doi.org/10.1016/j.scitotenv.2010.07.073.
14. Sakaguchi A, Steier P, Takahashi Y, Yamamoto M. Isotopic compositions of ^{236}U and Pu isotopes in "black substances" collected from roadsides in Fukushima Prefecture: fallout from the Fukushima Daiichi Nuclear Power Plant accident. Environ Sci Technol. 2014;48:3691–7. https://doi.org/10.1021/es405294s.
15. Yamamoto M, Komura K, Sakanoue M. ^{241}Am and Plutonium in Japanese rice-field surface soils. J Radiat Res. 1983;24:237–49. https://doi.org/10.1269/jrr.24.237.
16. Yamamoto M, Kuwabara J, Assinder DJ. Curium isotopes and americium-242m in Irish sea sediment. Radiochim Acta. 1998;83:121–6. https://doi.org/10.1524/ract.1998.83.3.121.
17. Yamamoto M, Oikawa S, Sakaguchi A, Tomita J, Hoshi M, Apsalikov KN. Determination of $^{240}Pu/^{239}Pu$ isotopic ratios in human tissues collected from areas around the Semipalatinsk Nuclear Test Site by sector-field high resolution ICP-MS. Health Phys. 2008;95:291–9. https://doi.org/10.1097/01.hp.0000314807.29873.71.
18. Yamamoto M, Sakaguchi A, Ochiai S, Imanaka T. Isotopic compositions of transuranic nuclides released by the Fukushima Daiichi Nuclear Power Plant accident: with emphasis on Cm isotopes. J Radioanal Nucl Chem. 2014;300:1045–52. https://doi.org/10.1007/s10967-014-3003-7.
19. Yamamoto M, Sakaguchi A, Ochiai S, Takada T, Hamataka K, Murakami T, Nagao S. Isotopic Pu, Am and Cm signatures in environmental samples contaminated by the Fukushima Daiichi Nuclear Power Plant accident. J Environ Radioact. 2014;132:31–46. https://doi.org/10.1016/j.jenvrad.2014.01.013.
20. Wang Z, Yang G, Zheng J, Cao L, Yu H, Zhu Y, et al. Effect of ashing temperature on accurate determination of plutonium in soil samples. Anal Chem. 2015;87:5511–5. https://doi.org/10.1021/acs.analchem.5b01472.
21. Bu W, Ni Y, Steinhauser G, Zheng W, Zheng J, Furuta N. The role of mass spectrometry for radioactive contamination assessment after the Fukushima nuclear accident. J Anal At Spectrom. 2018;33:519–46.
22. Zheng J, Tagami K, Watanabe Y, Uchida S, Aono T, Ishii N, Yoshida S, Kubota Y, Fuma S, Ihara S. Isotopic evidence of plutonium release into the environment from the Fukushima DNPP accident. Sci Rep. 2012;2:304. https://doi.org/10.1038/srep00304.
23. Schneider S, Walther C, Bister S, Schauer V, Christl M, Synal H-A, Shozugawa K, Steinhauser G. Plutonium release from Fukushima Daiichi fosters the need for more detailed investigations. Sci Rep. 2013;3:2988. https://doi.org/10.1038/srep02988.
24. Merz S, Steinhauser G, Hamada N. Anthropogenic radionuclides in Japanese food: environmental and legal implications. Environ Sci Technol. 2013;47:1248–56. https://doi.org/10.1021/es3037498.
25. Yamamoto M, Sakaguchi A. Environmental actinide elements associated with the Fukushima Daiichi Nuclear Power Plant accident. Chikyukagaku. 2015;49:173–84. (Japanese with English abstract).

Chapter 14
The Key Role of Isotopic Analysis in Tracing the Fukushima Nuclear Accident-Released Pu and Radiocesium Isotopes in the Environment

Youyi Ni, Jian Zheng, Qiuju Guo, and Hai Wang

Abstract The actinide plutonium (Pu) isotopes and the fission product radiocesium isotopes released in the Fukushima Daiichi Nuclear Power Plant (FDNPP) nuclear accident have drawn scientific attention in post-accident studies. In this chapter, studies that trace the Pu and radiocesium isotopes released from the FDNPP accident into the environment to ensure better nuclear emergency preparedness for the future were summarized. The characteristic $^{240}Pu/^{239}Pu$, $^{241}Pu/^{239}Pu$, and $^{135}Cs/^{137}Cs$ atom ratios of the FDNPP accident were determined to be 0.323–0.330, 0.128–0.135, and 0.333–0.343, respectively, which were distinct from those of global fallout. While Pu and radiocesium isotopic signatures from the accident were detected in the terrestrial environment, the release of Pu to the marine environment, if any, was negligible. And no data for ^{135}Cs in the marine environment has been reported yet.

Keywords Plutonium · Radiocesium · FDNPP · Atom ratio · Nuclear emergency

Y. Ni
National Institute of Radiological Sciences, National Institutes for Quantum and Radiological Science and Technology, Chiba, Japan

State Key Laboratory of Nuclear Physics and Technology, School of Physics, Peking University, Beijing, China

J. Zheng (✉)
National Institute of Radiological Sciences, National Institutes for Quantum and Radiological Science and Technology, Chiba, Japan
e-mail: zheng.jian@qst.go.jp

Q. Guo
State Key Laboratory of Nuclear Physics and Technology, School of Physics, Peking University, Beijing, China

H. Wang
National Institute of Radiological Sciences, National Institutes for Quantum and Radiological Science and Technology, Chiba, Japan

School of Nuclear Science and Technology, University of South China, Hengyang, China

G. Steinhauser et al. (eds.), *Nuclear Emergencies*,
Current Topics in Environmental Health and Preventive Medicine,
https://doi.org/10.1007/978-981-13-8327-4_14

14.1 Introduction

On March 11, 2011, equipment and structures of the Fukushima Daiichi Nuclear Power Plant (FDNPP) were damaged by a magnitude-9.0 earthquake and the earthquake-induced catastrophic tsunami. Since then, different amounts of radionuclides including plutonium (Pu) and radiocesium isotopes have been released into the environment. Tracing the FDNPP accident-released Pu and radiocesium isotopes and investigating their long-term fates in the environment have been important aspects of the post-accident studies for the scientific community. For the non-volatile Pu isotopes, because the released amount of the core inventory was small, recognizing the FDNPP accident-released Pu signature in the environment using conventional $^{239+240}$Pu activity is difficult. Since the atom ratios of Pu isotopes (e.g., ^{240}Pu/^{239}Pu, ^{241}Pu/^{239}Pu) vary with sources, such characteristic ratios can be reliable evidence for distinguishing the accident-released Pu in the environment. For radiocesium isotopes ^{134}Cs ($T_{1/2}$ = 2.06 y), ^{135}Cs ($T_{1/2} = 2.3 \times 10^6$ y), and ^{137}Cs ($T_{1/2}$ = 30.17 y), both ^{134}Cs and ^{137}Cs concentrations as well as ^{134}Cs/^{137}Cs activity ratios have been intensely monitored to identify the radioactivity contamination of the accident. However, due to technical difficulties in ^{135}Cs analysis, studies of the FDNPP released ^{135}Cs are rather scarce. Due to the short half-life of ^{134}Cs, the accident-released ^{134}Cs quickly decreases to the background level, making it impractical to tracing the long-term behaviors of this fission product in the environment. Therefore, it is of great scientific importance to obtain information on the release of ^{135}Cs and the ^{135}Cs/^{137}Cs isotopic fingerprint of the FDNPP accident for long-term studies.

The FDNPP consisted of six reactor units among which four (Units 1 to 4) were damaged to different extents. Units 1 to 3 were in operation at the time of the earthquake and a total of 256 tons of nuclear fuel had been loaded in these reactors [1, 2]. The Unit 4 reactor had been shut down before the accident for equipment replacement, and all the fuel in this reactor had been transferred to its spend fuel pool (SFP) in the reactor building. A total of 461 tons of nuclear fuel was stored in the SFPs of Units 1 to 4 [1, 2]. Therefore, during the accident, there were two possible sources of the released radionuclides, i.e., the reactor cores of Units 1 to 3 and the SFPs of Units 1 to 4. Thus, identifying the specific sources of Pu and radiocesium isotopes is instructive for accurate estimation of their release, and it also contributes to better understanding of the accident.

In this review, to highlight the key role of isotopic analysis in tracing the Fukushima nuclear accident-released Pu and radiocesium isotopes after they entered the environment: first, the background levels of Pu and radiocesium in the Japanese environment are summarized; then studies on characterizing the isotopic signatures of the FDNPP accident-released Pu and radiocesium isotopes in terrestrial environmental samples are introduced; possible Pu contamination in the marine environment is assessed; and last, some perspectives for future studies are discussed.

14.2 Background Levels of Pu and Radiocesium Isotopes in the Japanese Environment Before the Accident

Before the FDNPP accident, the main source of Pu and radiocesium in the Japanese environment was global fallout that was deposited following atmospheric nuclear weapon tests in the twentieth century. Due to the diversity of environmental conditions, the $^{239+240}Pu$ activity concentrations in Japanese soils (e.g., agricultural field soils and forest soils) were found to vary largely, typically ranging from 0.07 to 4.31 mBq/g [3–5]. In a recent work by Yang et al. [6], 80 archived surface soil samples (mainly school ground soils) collected in 1969–1977 from Fukushima and the adjacent prefectures were analyzed to establish baseline of Pu isotopes before the accident. The $^{239+240}Pu$ activity concentration in these surface soils ranged from 0.004 to 1.46 mBq/g, and the mean value of the $^{240}Pu/^{239}Pu$ atom ratio was 0.186 ± 0.015. The $^{240}Pu/^{239}Pu$ atom ratios in these samples were close to that of global fallout (0.18), indicating that global fallout was the major source. However, ^{241}Pu ($T_{1/2}$ = 14.4 y) in these surface soils was not detected because its concentration level was below the detection limit of the analytical method. To obtain the background of ^{241}Pu in these areas, Yang et al. employed the $^{241}Pu/^{239}Pu$ atom ratio obtained for a fallout reference material in Japan reported by Zhang et al. [7] to reconstruct the background distribution of ^{241}Pu in 1964 when the ^{241}Pu activity was presumably at the maximum. The ^{241}Pu activity concentration in these school ground surface soils was calculated to be 0.06–6.07 mBq/g and the mean $^{241}Pu/^{239+240}Pu$ activity ratio of 14.8 was obtained for the time of 1964.

Apart from supplementing the baseline of Pu in the terrestrial environment, Bu et al. [8] also made efforts to enrich the information about the Pu distribution in the marine environment before the accident. The concentrations of $^{239+240}Pu$ activity in the Japanese near-coast marine sediments ranged from 0.003 to 5.81 mBq/g, while the ^{241}Pu in these sediments was below the detection limit (2 mBq/g). In addition, the $^{240}Pu/^{239}Pu$ atom ratios ranged from 0.170 to 0.270, a high average ratio of 0.231 ± 0.025 ($n = 36$) in the sediments from the North Pacific side indicated a mixing of Pu from two endmembers with different $^{240}Pu/^{239}Pu$ atom ratios: the global fallout (0.18) and the Pacific Proving Grounds (PPG) close-in fallout Pu (0.30–0.36) [9].

Regarding the baseline of radiocesium isotopes in Japan, abundant reports on the ^{137}Cs activity concentration in the environment can be found in the literature. Because of the short half-life of ^{134}Cs and marginal yield in nuclear fission, its content in the Japanese environment was minute, and the $^{134}Cs/^{137}Cs$ activity ratio for the global-fallout background was approximately on the magnitude of 10^{-4} at the time of the accident [10]. Hampered by the difficulties in ^{135}Cs analysis, direct determination of the $^{135}Cs/^{137}Cs$ atom ratio in pre-accident environmental samples of Japan has not been reported yet. Yang et al. [10] analyzed the radiocesium isotopes (^{134}Cs, ^{135}Cs, and ^{137}Cs) in soil and plant samples collected in Fukushima

Prefecture after the accident. They reconstructed the $^{135}Cs/^{137}Cs$ atom ratios in the soil samples before the accident by considering the relative contribution of global fallout in these samples. Nevertheless, this estimation was rough since the contributions of global fallout in these contaminated samples were small (mostly <10%) and the calculated $^{135}Cs/^{137}Cs$ atom ratios for the global fallout showed large variation, ranging from 0.028 to 4.02 (decay corrected to March 11, 2011). Theoretically, because the fission yields of ^{135}Cs and ^{137}Cs from neutron fission of both ^{235}U and ^{239}Pu are similar, the productions of ^{135}Cs and ^{137}Cs from nuclear weapon tests are comparable and thus an isotopic ratio $^{135}Cs/^{137}Cs$ of around 1 can be expected for global fallout in the 1960s [11]. This has been supported by the measured $^{135}Cs/^{137}Cs$ atom ratios in the global-fallout contaminated sediment and rainwater samples reported in the literature [12, 13]. Therefore, after about a 50-year decay of ^{137}Cs, the characteristic $^{135}Cs/^{137}Cs$ atom ratio of the global fallout at the time of the accident (March, 2011) is presumably around 3. Zheng et al. [14] recently enlarged the $^{135}Cs/^{137}Cs$ atom ratio database for global fallout by analyzing a soil reference material (IAEA-soil-6) from Austria, the $^{135}Cs/^{137}Cs$ atom ratio in this global-fallout contaminated soil was calculated to be 2.58 ± 0.37 (decay corrected to March 11, 2011).

14.3 Characterizing the Signatures of Pu and Radiocesium Isotopes Released in the FDNPP Accident into the Terrestrial Environment

As discussed before, distinguishing the FDNPP accident-released Pu by using the $^{239+240}Pu$ activity concentration is difficult in the presence of global-fallout Pu since the release of non-volatile Pu from the accident was small. The isotopic compositions of Pu are more reliable tracers compared with the $^{239+240}Pu$ activity concentration. To clearly recognize the FDNPP accident-released Pu signature in the terrestrial environment, Zheng et al. [15] conducted Pu analysis with forest litter samples (S2 and S3 in Fig. 14.1b) collected two months after the accident. Because the soil-to-plant transfer factor of Pu is extremely low (10^{-3}–10^{-6}), the contribution of global-fallout Pu in these litter samples via soil-to-plant transfer was negligible (Fig. 14.1a). Thus the Pu present in these litter samples was derived exclusively from the FDNPP source. The $^{240}Pu/^{239}Pu$ (0.323–0.330) and $^{241}Pu/^{239}Pu$ (0.128–0.135, decay corrected to March 15, 2011) atom ratios in these forest litter samples were representative for the FDNPP-accident-source Pu and were distinctive from the background ($^{240}Pu/^{239}Pu$: 0.18, $^{241}Pu/^{239}Pu$: 0.0011, decay corrected to March 15, 2011) of global fallout, making it possible to trace the FDNPP-released Pu in the environment with these isotopic ratios. Slightly lower $^{240}Pu/^{239}Pu$ (0.303) and $^{241}Pu/^{239}Pu$ (0.103) atom ratios observed in a surface soil sample collected in the J-village facility (S6 in Fig. 14.1b) were believed to be the result of receiving a 13% contribution from the global fallout.

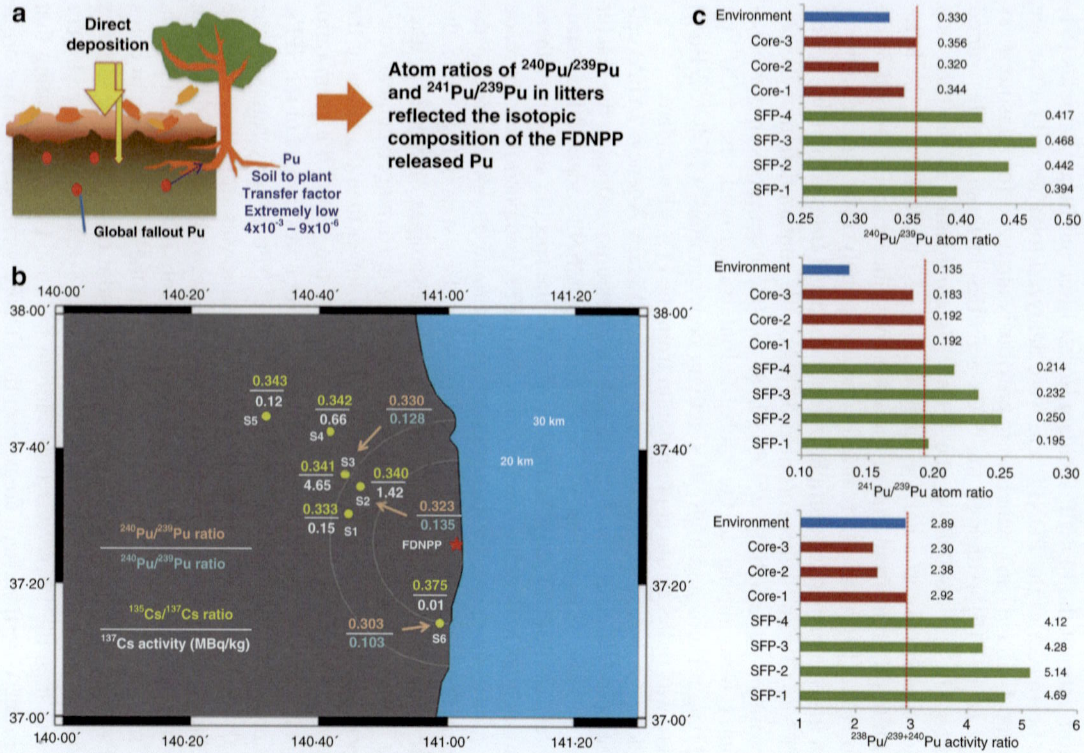

Fig. 14.1 Schematic diagram describing the presence of FDNPP-source Pu in the litter (**a**), environmental sampling sites for Pu and radiocesium analysis (**b**), and comparison of Pu atom ratios and activity ratios in environmental samples with those in reactor cores and SFPs (**c**). Figures were modified from Zheng et al. [16, 17]. Reprinted with permission from American Chemical Society, © 2013, 2014

For radiocesium isotopes, the $^{135}Cs/^{137}Cs$ atom ratio has been recognized as a new tracer for characterizing fallout from the nuclear power plant accident [17]. The $^{135}Cs/^{137}Cs$ atom ratios in nuclear reactors vary significantly with neutron fluxes and the reactor operation history, opening a new dimension to characterize a nuclear power plant accident with this distinctive ratio [18]. To obtain the characteristic $^{135}Cs/^{137}Cs$ atom ratio of the FDNPP accident, forest litter and lichen samples (S1-S5) collected northwest of the FDNPP site were analyzed together with a surface soil sample from J-village (S6), which were previously analyzed for Pu isotopes as shown in Fig. 14.1b [17]. Although the ^{137}Cs activities in these samples varied largely between 119 and 4649 Bq/g, the $^{135}Cs/^{137}Cs$ atom ratios in the litter and lichen samples varied only in a narrow range from 0.333 to 0.343, and the latter values were quite different from that of global fallout (theoretically around 3). In addition, Cao et al. [19] have also realized concurrent analysis of Pu and radiocesium isotopes in river suspended particles mainly collected from an area northwest of the FDNPP site to gain further information about the contamination of the terrestrial environment. Although $^{240}Pu/^{239}Pu$ (0.182–0.208) atom ratios in these suspended particles were similar to the global-fallout background, the $^{135}Cs/^{137}Cs$ (0.329–0.391) atom ratios and high ^{137}Cs activity concentrations (23.4–152 Bq/g) suggested there was radiocesium contamination from the FDNPP accident.

After the accident, researchers of the Japan Atomic Energy Agency (JAEA) calculated the isotopic compositions of the Pu and radiocesium isotopes in the reactor cores of Units 1–3 and SFPs of Units 1–4 using the ORIGEN2 code and fuel burn-up data provided by TEPCO [20]. To further identify the specific sources of these released radionuclides among the reactor cores and SFPs, comparisons were made between the model-simulated Pu and radiocesium isotopic compositions and the actually measured values in the environmental samples. As is seen in Fig. 14.1c, the highest $^{240}Pu/^{239}Pu$ atom ratio among the environmental samples obtained for litter collected in a Fukushima forest was close to the ratios of the damaged nuclear reactors, but the value of this highest ratio was significantly lower than the ratios of the SFPs. The same phenomenon was found in the case of the $^{241}Pu/^{239}Pu$ atom ratio, indicating that Pu was released from the damaged reactor cores instead of the SFPs. This viewpoint was further supported by the $^{238}Pu/^{239+240}Pu$ activity ratios in the environmental samples among which the maximum value was still lower than the maximum values in the SFPs. A similar conclusion was drawn by using radiocesium tracer. In the litter and lichen samples receiving exclusively FDNPP contamination, the $^{135}Cs/^{137}Cs$ atom ratios were more similar to those in the damaged reactor cores than the SFPs (Fig. 14.2a). Furthermore, $^{135}Cs/^{137}Cs$ atom ratios in these samples corresponded very well with those in the Units 2 and 3 reactor cores, and were somewhat lower than the $^{135}Cs/^{137}Cs$ atom ratio in the Unit 1 core, indicating that radiocesium in these northwestern areas from the FDNPP was likely to have originated from the reactor cores of Units 2 and 3. The slightly higher $^{135}Cs/^{137}Cs$ atom ratio in the surface soil from J-village (S6) was likely to have resulted from the contribution of the Unit 1 core and/or the mixing of global-fallout radiocesium with the FDNPP-accident-derived radiocesium. In addition, a cluster map illustrating the $^{240}Pu/^{239}Pu$ and the $^{135}Cs/^{137}Cs$ atom ratios in the reactor cores, SFPs and the heavily

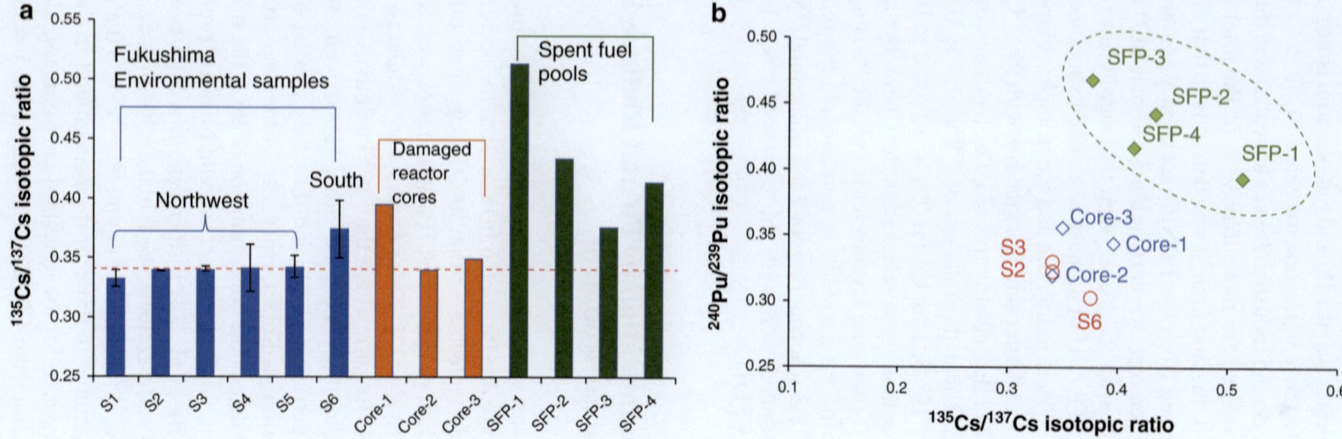

Fig. 14.2 Comparison of the $^{135}Cs/^{137}Cs$ atom ratios in environmental samples with those of the reactor cores and SFPs (**a**), and cluster map of $^{240}Pu/^{239}Pu$ and $^{135}Cs/^{137}Cs$ for source identification (**b**). Figures were modified from Zheng et al. [17]. Reprinted with permission from American Chemical Society, © 2014

contaminated environmental samples (S2, S3: litter, S6: soil) is presented in Fig. 14.2b. It clearly illustrates that the Pu and radiocesium isotopic compositions in the samples were closer to those in the reactor core of Unit 2. Considering that the Ministry of Economy, Trade and Industry (Japan) has estimated the released amount of ^{137}Cs from Unit 2 accounted for more than 93% of the total releases from the three damaged reactors, for the first time, it was determined that the major source of Pu and radiocesium released from the FDNPP accident was the damaged reactor core in the Unit 2 [17]. Therefore, the early release estimate of ^{137}Cs, i.e., 36.6 PBq made by Stohl et al. [21], with the assumption that significant releases of fission products occurred in the SFP of Unit 4, must be an overestimation.

Based on the ^{137}Cs/$^{239+240}$Pu activity ratio and the isotopic compositions of Pu and Cs in the heavily contaminated litter samples together with the ^{137}Cs releases, Zheng et al. [16–15] estimated the released amounts of Pu isotopes and the ^{135}Cs during the accident. It was concluded that the releases of ^{238}Pu (2.9×10^9–6.9×10^9 Bq), $^{239+240}$Pu (1.0×10^9–2.4×10^9 Bq), and ^{241}Pu (1.1×10^{11}–2.6×10^{11} Bq) were four orders of magnitude lower than those released in the Chernobyl accident, and only a minute percent (10^{-5}%) of the core inventories for all the Pu isotopes was released into the environment during the accident. In comparison, the release of ^{135}Cs was estimated to be 6.7×10^{-5} PBq, accounting for 2.0% of the total ^{135}Cs core inventory. This conclusion corresponded well with the estimated ^{137}Cs percent of release (~2%) made by the IAEA [22].

14.4 Possible Contamination by Pu and Radiocesium in the Marine Environment

After the accident, seawater and sediment samples in the Northwest Pacific Ocean were analyzed to detect possible Pu contamination from the accident [8, 15, 23–26]. For the seawater samples collected in the 30–200 km zone off the FDNPP, the $^{239+240}$Pu activity concentrations and ^{240}Pu/^{239}Pu atom ratios were 0.43–5.59 mBq/m^3 and 0.227–0.284, respectively [27]. These results were similar to those in seawater of the distant sea (446–1316 km), where the $^{239+240}$Pu activity concentrations were 1.21–2.19 mBq/m^3, while the ^{240}Pu/^{239}Pu atom ratios were 0.198–0.322 [25]. Plutonium isotopes are particle-reactive and thus they can be quickly scavenged from seawater to the bottom sediment especially in the near-coast marginal seas. Therefore, if there is any detectable contamination of Pu in the marine environment from the FDNPP accident, the Pu signatures should be likely presented in the sediment. Considering this, Bu et al. [24] further determined Pu isotopes in sediment samples from the North Pacific collected both in the <30 km zone and in the >30 km zone off the FDNPP. The $^{239+240}$Pu activities and the ^{240}Pu/^{239}Pu atom ratios in these sediments were all comparable to those of the background in the near-coast sediments. Compared with the case for the terrestrial environment, identifying the possible Pu contamination from the FDNPP accident in the marine environment is more difficult because of the dilution effect of seawater and the fact that the released amount of Pu was small. More importantly, the similarity of the characteristic $^{240}Pu/^{239}Pu$ atom ratios for the FDNPP source (0.323–0.330) and the PPG close-in fallout Pu (0.30–0.36) make it difficult to distinguish the contribution of FDNPP in seawater and sediments with this single tracer. To provide more solid evidence on tracing the possible Pu contamination in the marine environment, the $^{241}Pu/^{239}Pu$ atom ratios in the sediments collected both inside and outside the 30 km zone after the accident were determined. The $^{241}Pu/^{239}Pu$ atom ratios in all of the sediments ranged narrowly from 0.0012 to 0.0016, which corresponded well with those of global fallout (0.0011) and the PPG close-in fallout (0.0018–0.0025), but they were distinctly different (about 100 times lower) from that of the FDNPP source (0.128–0.135). Therefore, it was concluded that the release of Pu isotopes from the FDNPP to the marine environment, if any, was negligible. Regarding the radiocesium isotopes, it was estimated that around 80% of the atmospheric release of ^{137}Cs was eventually deposited over the west North Pacific, and additional radioactive waste water has been directly discharged or leaked into the Pacific Ocean [28, 29]. However, no data for the $^{135}Cs/^{137}Cs$ atom ratios in the marine environment after the accident has been reported yet.

14.5 Perspectives for Future Study

In the post-accident studies reviewed here, efforts have been made to trace the FDNPP-source radionuclides and to reveal their temporal variations and spatial distributions in the environment. Regarding the Pu and radiocesium isotopes in the environment, the authors of this review suggest the following points for future study:

1. Characterizing the Pu and radiocesium isotopic compositions in highly radioactive microparticles (hot particles) released from the FDNPP accident.
2. Long-term monitoring of the variations of Pu and radiocesium isotopes in the North Pacific.
3. Studying the riverine transport of Pu and radiocesium isotopes with soil particles by land runoffs or floods.
4. Investigating the vertical migration of Pu and radiocesium isotopes in soil in the heavily contaminated areas.
5. Assessing the bioavailability of FDNPP accident-released Pu and radiocesium isotopes to plants and animals.

Acknowledgements This work was supported by the Grant of Fukushima Prefecture related to Research and Development in Radiological Sciences, the Grand-in-Aid for Scientific Research by the Ministry of Education, Culture, Sports, Science and Technology, Japan (Grant number JP17k00537), the Kakenhi Grant-in-Aid for Scientific Research on Innovative Areas (24110004), the Sumitomo Environmental Foundation, and the Agency for Natural Resources and Energy (METI), Japan. Youyi Ni thanks the China Scholarship Council for offering a scholarship (201606010056) for supporting his Ph.D. study.

References

1. Grambow B, Poinssot C. Interactions between nuclear fuel and water at the Fukushima Daiichi reactors. Elements. 2012;8:213–9.
2. Schwantes JM, Orton CR, Clark RA. Analysis of a nuclear accident: fission and activation product releases from the Fukushima Daiichi nuclear facility as remote indicators of source identification, extent of release, and state of damaged spent nuclear fuel. Environ Sci Technol. 2012;46:8621–7.
3. Muramatsu Y, Yoshida S, Tanaka A. Determination of Pu concentration and its isotope ratio in Japanese soils by HR-ICP-MS. J Radioanal Nucl Chem. 2003;255:477–80.
4. Ohtsuka Y, Iyogi T, Kakiuchi H, et al. Evaluation of $^{239+240}$Pu, ^{137}Cs and natural ^{210}Pb fallout in agricultural upland fields in Rokkasho, Japan. J Radioanal Nucl Chem. 2004;261:625–30.
5. Yamamoto M, Komura K, Sakanoue M. ^{241}Am and plutonium in Japanese rice-field surface soils. J Radiat Res. 1983;24:237–49.
6. Yang G, Zheng J, Tagami K, et al. Plutonium concentration and isotopic ratio in soil samples from central-eastern Japan collected around the 1970s. Sci Rep. 2015;5:9636.
7. Zhang Y, Zheng J, Yamada M, et al. Characterization of Pu concentration and its isotopic composition in a reference fallout material. Sci Total Environ. 2010;408:1139–44.
8. Bu W, Zheng J, Aono T, et al. Vertical distributions of plutonium isotopes in marine sediment cores off the Fukushima coast after the Fukushima Dai-ichi Nuclear Power Plant accident. Biogeosciences. 2013;10:2497–511.
9. Buesseler KO. The isotopic signature of fallout plutonium in the North Pacific. J Environ Radioact. 1997;36:69–83.
10. Yang G, Tazoe H, Yamada M. ^{135}Cs activity and ^{135}Cs/^{137}Cs atom ratio in environmental samples before and after the Fukushima Daiichi Nuclear Power Plant accident. Sci Rep. 2016;6:24119.
11. Snyder DC, Delmore JE, Tranter T, et al. Radioactive cesium isotope ratios as a tool for determining dispersal and re-dispersal mechanisms downwind from the Nevada Nuclear Security Site. J Environ Radioact. 2012;110:46–52.
12. Chen HW, Lee T, Ku TL, et al. Production ratio of nuclear fallout ^{137}Cs/^{135}Cs. Chin J Physiol. 2008;46:560–9.
13. Lee T, Teh-Lung K, Hsiao-Ling L, et al. First detection of fallout Cs-135 and potential applications of ^{137}Cs/^{135}Cs ratios. Geochim Cosmochim Acta. 1993;57:3493–7.
14. Zheng J, Cao L, Tagami K, et al. Triple-quadrupole inductively coupled plasma-mass spectrometry with a high-efficiency sample introduction system for ultratrace determination of ^{135}Cs and ^{137}Cs in environmental samples at femtogram levels. Anal Chem. 2016;88:8772–9.
15. Zheng J, Tagami K, Watanabe Y, et al. Isotopic evidence of plutonium release into the environment from the Fukushima DNPP accident. Sci Rep. 2012;2:304.
16. Zheng J, Tagami K, Uchida S. Release of plutonium isotopes into the environment from the Fukushima Daiichi nuclear power plant accident: what is known and what needs to be known. Environ Sci Technol. 2013;47:9584–95.
17. Zheng J, Tagami K, Bu W, et al. ^{135}Cs/^{137}Cs isotopic ratio as a new tracer of radiocesium released from the Fukushima nuclear accident. Environ Sci Technol. 2014;48:5433–8.
18. Bu W, Ni Y, Steinhauser G, et al. The role of mass spectrometry in radioactive contamination assessment after the Fukushima nuclear accident. J Anal At Spectrom. 2018;33:519–46.
19. Cao L, Zheng J, Tsukada H, et al. Simultaneous determination of radiocesium (^{135}Cs, ^{137}Cs) and plutonium (^{239}Pu, ^{240}Pu) isotopes in river suspended particles by ICP-MS/MS and SF-ICP-MS. Talanta. 2016;159:55–63.
20. Nishihara K, Iwamoto H, Suyama K Estimation of fuel compositions in Fukushima-Daiichi nuclear power plant. JAEA-Data/Code 2012, No. 2012-018; 2012.
21. Stohl A, Seibert P, Wotawa G, et al. Xenon-133 and caesium-137 releases into the atmosphere from the Fukushima Dai-ichi nuclear power plant: determination of the source term, atmospheric dispersion, and deposition. Atmos Chem Phy. 2012;12:2313–43.

22. International Atomic Energy Agency (IAEA). Report of Japanese Government to the IAEA Ministerial conference on nuclear safety: the accident at TEPCO's Fukushima nuclear power stations. Nuclear Emergency Response Headquarters Government of Japan; 2011.
23. Bu W, Zheng J, Guo Q, et al. A method of measurement of ^{239}Pu, ^{240}Pu, ^{241}Pu in high U content marine sediments by sector field ICP–MS and its application to Fukushima sediment samples. Environ Sci Technol. 2013;48:534–41.
24. Bu W, Fukuda M, Zheng J, et al. Release of Pu isotopes from the Fukushima Daiichi Nuclear Power Plant accident to the marine environment was negligible. Environ Sci Technol. 2014;48:9070–8.
25. Men W, Zheng J, Wang H, et al. Establishing rapid analysis of Pu isotopes in seawater to study the impact of Fukushima nuclear accident in the Northwest Pacific. Sci Rep. 2018;8:1892.
26. Zheng J, Aono T, Uchida S, et al. Distribution of Pu isotopes in marine sediments in the Pacific 30 km off Fukushima after the Fukushima Daiichi nuclear power plant accident. Geochem J. 2012;46:361–9.
27. Bu W, Zheng J, Guo Q, et al. Ultra-trace plutonium determination in small volume seawater by sector field inductively coupled plasma mass spectrometry with application to Fukushima seawater samples. J Chromatogr A. 2014;1337:171–8.
28. Kanda J. Continuing ^{137}Cs release to the sea from the Fukushima Dai-ichi Nuclear Power Plant through 2012. Biogeosciences. 2013;10:6107–13.
29. Tsumune D, Tsubono T, Aoyama M, et al. Distribution of oceanic ^{137}Cs from the Fukushima Dai-ichi Nuclear Power Plant simulated numerically by a regional ocean model. J Environ Radioact. 2012;111:100–8.

Chapter 15
Radioiodine Releases in Nuclear Emergency Scenarios

Olivier Masson, Jochen Tschiersch, Luke S. Lebel, Herbert Wershofen, Jerzy Wojciech Mietelski, Georg Steinhauser, Éric Blanchardon, Laurent Cantrel, Anne-Cécile Grégoire, and Denis Quélo

Abstract This document provides a comprehensive overview study on the physico-chemical speciation of radioiodine observed in the atmosphere after various emissions related to nuclear activities: nuclear weapon tests, accident and incident releases, and routine discharges. The study covers different types of nuclear facilities including medical isotope production facilities (MIPFs), reprocessing plants (RPs), and nuclear power plants (NPPs). Most attention is paid to ^{131}I which has a major human health impact in the early stages of a nuclear emergency situation with regard to inhalation. Iodine-131 combines a high yield by neutron-induced nuclear fission of ^{235}U (2.87%) or ^{239}Pu (3.8%), high dose coefficients, and a radioactive half-life long enough to allow for spreading at global scales and entering the food chain but sufficiently short to produce a significant dose commitment when inhaled or ingested. Reliable dose assessment requires both detailed and valid information on the physico-chemical composition of ^{131}I present in the air. Apart from reactor explosions and fires, which produce large amounts of particles and may therefore

---

O. Masson (✉) · E. Blanchardon · L. Cantrel · A.-C. Grégoire · D. Quélo
Institut de Radioprotection et de Sûreté Nucléaire (IRSN), Fontenay-aux-Roses, France
e-mail: olivier.masson@irsn.fr

J. Tschiersch
Helmholtz Zentrum München GmbH (HMGU), Institute of Radiation Protection, Neuherberg, Germany

L. S. Lebel
Chalk River Laboratories, Canadian Nuclear Laboratories (CNL), Chalk River, ON, Canada

H. Wershofen
Physikalisch-Technische Bundesanstalt (PTB), Braunschweig, Germany

J. W. Mietelski
The Henryk Niewodniczański Institute of Nuclear Physics, Polish Academy of Sciences, Kraków, Poland

G. Steinhauser
Leibniz Universität Hannover, Institute of Radioecology and Radiation Protection, Hannover, Germany

G. Steinhauser et al. (eds.), *Nuclear Emergencies*,
Current Topics in Environmental Health and Preventive Medicine,
https://doi.org/10.1007/978-981-13-8327-4_15

favor the presence of iodine in particulate form at short distance, other nuclear accident scenarios will lead fairly rapidly to a dominant gaseous radioiodine proportion in the atmosphere.

Keywords Iodine · Radioiodine · Nuclear accident · Thyroid · Internal exposure · Radiation protection · Environmental release · Environmental monitoring · Routine discharges

15.1 Introduction

The goal of this chapter is to provide basic and updated knowledge about radioiodine (i.e., radioactive isotopes of iodine) speciation in the atmosphere and to provide information about the way to refine inhalation dose assessment and reduce associated uncertainties. Within this framework, different emission scenarios covering a wide range of nuclear activities and release situations are examined. Without neglecting the contribution of the ^{131}I dose induced by ingestion, that may rapidly overwhelm that of inhalation if proper ban on sales or on foodstuff consumption are not taken rapidly after the deposition of airborne contaminant, this review focusses on the contamination of the atmosphere and the dose induced by inhalation.

Radioiodine is of major concern for public health when released in large quantities into the atmosphere, as during nuclear accidents or even significant nuclear incidents at short distances. Whether radioactive or not, iodine is necessary for the thyroid gland to synthesize hormones involved in the human body metabolism. It is considered as an essential element. Especially in areas of shortage in iodine supply in the daily diet, the human body is seeking for each iodine molecule, ion, or compound either in the breathing air or in the food. Among the various population ages, children are most sensitive to iodine deficiency and to the uptake of radioactive iodine.

The chemistry of iodine-containing species is complex, and their chemical existence once released into the atmosphere will depend on many interactions with other atmospheric compounds and will vary according to solar radiation. Iodine is a volatile element with high vapor pressures, even at relatively low temperatures. Molecular iodine (I_2) sublimates quite easily even at ambient temperature and boils at 184 °C owing to this high vapor pressure. In addition, organically-bound iodine species are even more volatile, where, for example, methyl iodide (CH_3I) boils at 42 °C. Particulate forms of iodine are often salt or oxide forms (e.g., CsI or I_xO_y such as I_2O_3), and iodine coexists both in these gaseous and particulate forms in the atmosphere. Even if it is expected to remain mostly in gaseous species (organic and inorganic), parts of them will condensate and nucleate to form nanoparticles, or adsorb onto atmospheric aerosols. However, the transfer kinetics related to adsorption on particles, the kinetics of the reactions with other airborne compounds and finally the kinetics of the conversion into particulate iodine by condensation and heterogeneous nucleation are not yet fully understood and are also a matter of uncertainties and biases in inhalation dose assessment.

Being a member of the halogen group, iodine is a reactive element and has a complex chemistry with other atmospheric compounds [1]: aerosols, ozone, H_xO_y, nitrogen oxides (NOx), volatile and semi-volatile organic compounds (VOC, SVOC), etc. In addition, these interactions are driven by light and UV radiation to form gaseous organic iodides (CH_3I, CH_2I_2, $CH_3CH_2CH_2I$, etc.) during diurnal conditions while they produce mostly inorganic gaseous species (I_2, HI, IO, HOI, HIO_3, $IONO_2$, etc.) in night conditions [2, 3], in addition to aerosol-bound species (mixed iodinated aerosols, iodide salts, and iodine oxide aerosols [3] formed by reaction with ozone to give IO_2, I_2O_3, I_2O_4, I_2O_5, etc. Gaseous iodine species such as CH_3I and even more I_2 are sensitive to photo-dissociation (or photolysis) under solar UV radiation. The photolysis of I_2 and CH_3I generate active atomic iodine (I) that will react with the previously mentioned atmospheric compounds and will be recycled through numerous intermediates in both gaseous and aerosol phases over time scales of hours to days [4]. The main final CH_3I photolysis products in the air are in the form of I_xO_y [5]. As a consequence of photo-dissociation, the concentrations of I_2 and CH_3I and that of their reactions products will rapidly change between daytime and nighttime [6, 7]. According to kinetic parameters used in iodine behavior models, it has been found that I_2 has a short lifetime of about 20 s before photolysis in a sunlit atmosphere [8] or that 95% of I_2 released during daytime will be photo-dissociated in less than 1 min. Over a period of a few hours [9], a rapid evolution of the physico-chemical forms of iodine can be observed as a function of seasonal parameters with an almost total conversion of released I_2 into INOx in winter conditions and INOx plus organic iodine in summer conditions where photolysis plays a predominant role. Based on kinetic parameters related to photo-dissociation rates, I_2 is thus theoretically not expected to be found during daytime. However, iodine speciation data acquired after nuclear accidents demonstrate that gaseous inorganic iodine exists in a rather high proportion after a long range transport. Relative comparison can be made with CH_3I [10] whose photo-dissociation requires between 1.1 and 4 days [6] and even 8 days (CH_3I is photolyzed about 600 times more slowly than I_2) [8]. Conversely, enhanced oceanic emissions of HOI (hypoiodous acid) and I_2 happen during nighttime in addition to the increase of INOx as demonstrated when using kinetic parameter-based theoretical calculations [6, 7]. Filistovic and Nedveckaite [6] suggested that the atmospheric persistence of inorganic iodine species in a rather significant proportion, as observed after nuclear accidents, is less linked to I_2 than to other inorganic species like HOI or $IONO_2$ species. The global chemical complexity of iodine also encompasses iodinated reaction products which in turn will behave in a different way in the atmosphere depending on their physico-chemical features, UV radiation, and the mechanisms involved in the contamination of the environment: gas-to-particle conversion and transfer from one species to another, deposition and interception, biomass integration, re-emission of aerosol-bound iodine (liquid or solid), or gaseous re-volatilization. All the different chemical and physical forms that radioiodine (i.e., radioactive isotopes of iodine) can take in the atmosphere will influence their individual behavior, and thus their persistence in the atmosphere and the contamination of other compartments of the biosphere. The same statement stands for incident releases, and routine discharges. By extrapolation, there may ultimately be consequences for long-range deposition or transport of radioiodine with dose–response implications received in proportion to inhaled organic, inorganic, and particulate species, as well as levels of contamination of marine and continental ecosystems. Therefore, a comprehensive quantification of the exposure to radioiodine is of highest importance for radioprotection issues.

15.2 Synopsis of Radioiodine Presence in the Atmosphere

Since the beginning of the nuclear era, there has been a large variety of radioactive iodine emissions to the environment. Apart from ^{127}I, which is the only stable isotope, there exist 42 radioactive iodine isotopes or isomers with atomic numbers between 108 and 145, and 13 of these are fission products. Thirteen iodine isotopes have a half-life longer than 1 h, and four have a half-life ranging from a few days to about 60 days (Table 15.1). Radioactive iodine isotopes are generated

Table 15.1 Radioiodine half-lives longer than 10 min and main γ–X-ray energies (keV) and emission intensity (%)

Radioiodine mass number	Half-life	Specific activity (Bq/g)	Main γ–X-ray energy (keV), emission intensity (%)	Other analytically relevant γ–X-rays	Maximum β emission (keV), emission intensity (%)
119	19.3 min				2230 (46%) 3252 (38%)
120	1.35 h	$7.15{\cdot}10^{17}$			
120 m	53 min				
121	2.12 h	$4.5{\cdot}10^{17}$			2058 (76.6%)
123	13.27 h	$7.2{\cdot}10^{16}$	159 (83%)		
124	4.18 days		602.7 (63%)	723 (10%), 1691 (11%)	1532 (11.7%)
125	59.4 days	$6.4{\cdot}10^{14}$	35.5 (6.68%)	27.2 (40%), 27.5 (76%)	
126	13.11 days	$7.1{\cdot}10^{16}$	338.6 (34%)	666.3 (33%)	869.0 (33.4%)
128	24.99 min	$2.2{\cdot}10^{18}$	442.9 (17%)		2119.0 (76.7%)
129	$1.57{\cdot}10^{7}$ years	$6.5{\cdot}10^{6}$	39.6 (7.5%)	29.5 (20%), 29.8 (38%)	151.2 (99.5%)
130	12.36 h	$7.2{\cdot}10^{16}$	536 (99%)	668.5 (96%), 739.5 (82%)	1004.9 (48%)
131	8.02 days	$4.6{\cdot}10^{15}$	364.5 (82%)	637.0 (7.1%)	606.3 (89.4%)
132	2.28 h	$3.8{\cdot}10^{17}$	667.7 (99%)	772.6 (76%)	2140.0 (19%) 1185.0 (18.8%)
132 m	1.38 h	$6.3{\cdot}10^{17}$	600 (14%)	173.7 (8.8%)	1482.9 (8.6%)
133	20.9 h	$4.2{\cdot}10^{16}$	529.9 (87%)		1227.0 (83.4%)
134	52.6 min	$9.9{\cdot}10^{17}$	847 (95%)	884.0 (65%)	1307.6 (30.4%)
135	6.61 h	$1.3{\cdot}10^{17}$	1260 (29%)	1131.5 (22.6%)	1387.6 (23.8%)

Compilation of data from [11–14]

during the nuclear fuel life cycle (from spontaneous fission of uranium in soil to energy production and reprocessing of spent nuclear fuel) or by various techniques involved in radiopharmaceutical production like neutron bombardment. Among the different released and transported radioactive iodine isotopes in the atmosphere, most attention is paid to ^{131}I. Iodine-131 combines a high yield by (thermal) nuclear fission of ^{235}U (2.88%) or ^{239}Pu (3.8%), high dose coefficients, and a radioactive half-life long enough to let it spread at worldwide scale and enter the food chain but also sufficiently short to produce a high dose commitment when inhaled or ingested. Major decay of ^{131}I into ^{131}Xe takes place through β and γ emissions with a short half-life of 8.02 days. The thermal fission of ^{235}U also produces another iodine isotope of interest, namely ^{129}I with a ratio ^{131}I/^{129}I = 3.59 [15] according to the most recent 2017 OECD "Joint Evaluated Fission and Fusion File" (JEFF-3.3). Only ^{129}I has a very long half-life ($T_{1/2} = 1.57 \cdot 10^7$ years), and is also the only radioactive iodine isotope to be naturally produced by nuclear reactions of atmospheric Xe under the impact of cosmic radiation [16], and to a lesser extent in soil by spontaneous fission of ^{238}U, thermal neutron-induced fission of ^{235}U (fission yield 0.9%), or by neutron activation reactions of tellurium isotopes in soil. However, its natural inventory estimated in total to ~ 230 kg (ca. 1.5 TBq) is largely overwhelmed by anthropogenic emissions from nuclear activities [2, 16].

15.3 Physiological Aspects and Health Impact

Iodinated compounds exhibit a high affinity with different organs (mainly the thyroid gland) whose metabolism is essential for the development and functioning of the body. Iodine is an essential component of the thyroid hormones thyroxine (T4) and triiodothyronine (T3), which regulate metabolic processes and are critical to growth and development. The iodine concentration in the thyroid of an adult is about 500 μg g^{-1}, and 80–90% of iodine present in the body of an adult is contained in its thyroid. The World Health Organization (WHO) recommends a daily iodine intake of 150 μg for an adult (and 200 μg for a pregnant or breastfeeding woman). Large regional variations are observed, from less than 20 μg days^{-1} to more than 10,000 μg days^{-1}. Inhalation of air contaminated with radioiodine, i.e., during the plume presence, or ingestion of contaminated foodstuff within days or even a few weeks after deposition lead to internal irradiation of the thyroid gland and increases the risk of thyroid cancer. WHO-recommended daily iodine intakes converted into gaseous-only $^{131}I_2$ and $CH_3{}^{131}I$ activities would be equivalent to about 0.69 and 0.35 TBq, respectively. Thus the mass of radioiodine that could be incorporated by an individual, as a consequence of a nuclear accident, would be much lower than his stable iodine intake. The thyroid will readily absorb 5–90% of the incorporated radioiodine, depending on the dietary intake of stable iodine, with highest thyroid uptake in the case of a stable iodine deficiency. Most of the remaining radioiodine intake is quickly excreted in urine. The thyroid gland is a small organ of about 15 cm^3–20 cm^3 only, for an adult. When inhaled or ingested, ^{131}I will thus concentrate in this gland where a relatively high specific dose will be delivered over about one month, mainly through β^- radiation of mean energy 182 keV with 100% yield and maximum emission at 606 keV (90% intensity) prone to irradiate thyroid tissues up to 2 mm range [17].

In comparison, ^{129}I decay emits β^- particles with a maximum energy of 151 keV, 100% (mean 37 keV). Because of the low levels of ^{129}I activity produced in nuclear reactors and released in the environment in case of an accident (and its low specific activity of $6.5 \cdot 10^6$ $Bq \cdot g^{-1}$ in general), the dose induced by ^{129}I exposure remains negligible compared to that from ^{131}I. However, because of its very long half-life, the study of ^{129}I is a matter of interest for ^{131}I dose reconstruction and in the framework of radiological waste disposal. Indeed, in a long-term basis after the closure of a repository ^{129}I as well as ^{36}Cl, ^{79}Se, and ^{99}Tc are expected to be a major dose source for humans due to their high mobility in the environment and anionic nature. In the safety assessment of the spent nuclear fuel in a long-term basis, ^{129}I and ^{36}Cl are classified as the first (top) priority radionuclides [18].

Following the Chernobyl accident (April 1986), radio-induced hypothyroidism appeared for high dose exposures (about 10 Gy and above), while radio-induced thyroid cancer incidence increased about 3–5 years [19] after exposure among children to doses higher than 100 mGy (Fig. 15.1a) [21–23]. The younger the age at exposure, the higher the risk (Fig. 15.1b). After this accident, it has been found that children having stable iodine deficiency experienced at least twice as high thyroid exposure as compared with children in the case of a dietary iodine sufficiency. An increase of cancer risk with thyroid dose was also reported among "Chernobyl liquidators" exposed at adult age [24].

Iodine-containing species in the atmosphere are usually categorized into gaseous inorganic species with I_2 (elemental iodine, also called molecular iodine) as the representative species; gaseous organic species with CH_3I (methyl iodide) as the representative, and particulate iodine (aerosol-bound). Once inhaled, these species will deposit in different proportions in the respiratory tract and integrate partly or almost totally the thyroid gland. Biokinetic effects are integrated in a single parameter named dose coefficient (dose per unit of intake, expressed in sievert per incorporated becquerel of radionuclide). Dose coefficients are published by the International Commission on Radiological Protection (ICRP) for the three representative iodine-containing species (Table 15.2). According to Morgan et al. [27], the average CH_3I deposition in the respiratory tract is 72% (53–92%) of intake [27]. The deposition of inhaled I_2 is even higher (>90%). For both vapors, the absorption of deposited activity to blood is complete and very rapid. About a third of the activity transferred to blood concentrates in the thyroid within 24 h. The deposition of particulate iodine depends on the size distribution of the aerosol. Gaseous organic and inorganic iodine species have higher inhalation dose coefficients than aerosol-bound because of their larger deposition in the respiratory tract: For elemental iodine vapor (I_2), it is assumed that 100% are deposited in the respiratory tract [28], while regarding the aerosol-bound fraction, about half of the aerosol with an activity median aerosol diameter (AMAD) of 1 μm will deposit. Pathways of particle clearance rely on exhalation, dissolution in the lung fluids, transport in the alimen-

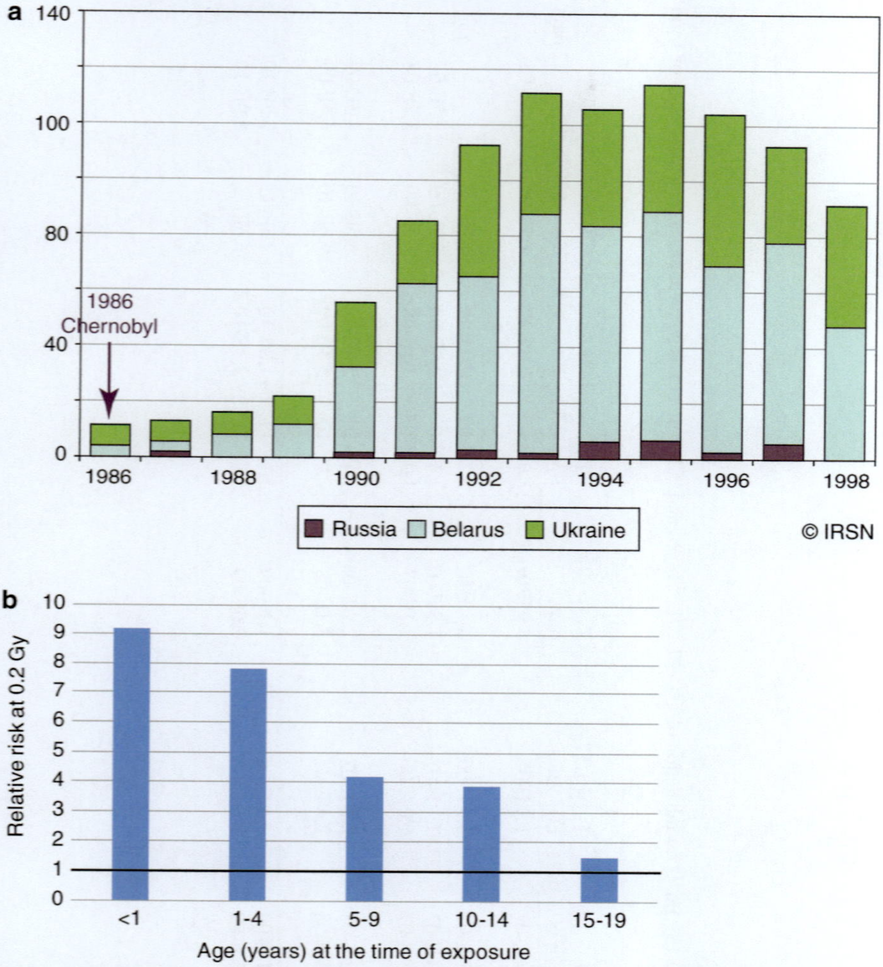

Fig. 15.1 (**a**) Temporal change of the yearly thyroid cancer number in children aged <15 when exposed to the Chernobyl accident. Reprinted with permission, © IRSN. (**b**) Relative Risk (RR) of thyroid cancer according to the age at exposure to the Chernobyl accident, after [20]

tary tract, and absorption in blood. For the particulate fraction, dose coefficients vary with the dissolution rate of the deposited particulate species in the lung fluids: reference absorption types fast, moderate, or slow (F, M, S) are considered by the ICRP. Type F (rapid and complete absorption) is recommended by default for ^{131}I when no specific information is available for members of the public [25]. The fraction of an ingested activity that is absorbed into blood is quantified according to f_1 values; 1 corresponding to complete absorption is recommended for ingested iodine.

Table 15.2 Thyroid equivalent dose coefficients (Sv Bq^{-1}) for inhalation of ^{125}I, ^{129}I, ^{131}I, ^{132}I, ^{133}I, and ^{132}T compounds by an adult, with aerosol sizes of 1 µm [25] and 5 µm [26]

	^{125}I $T_{1/2}$ = 60.1 days	^{129}I $T_{1/2}$ = 1.57 E07 years	^{131}I $T_{1/2}$ = 8.04 days	^{132}I $T_{1/2}$ = 2.3 h	^{133}I $T_{1/2}$ = 20.8 h	^{135}I $T_{1/2}$ = 6.61 h	^{132}Te $T_{1/2}$ = 3.2 days
Elemental iodine (I_2) and unspecified forms	2.7 E-07	1.9 E-06	3.9 E-07	3.6 E-09	7.6 E-08	1.5E-08	7.6E-08
Methyl iodide (CH_3I) ethyl iodide (C_2H_5I)	2.1 E-07	1.5 E-06	3.1 E-07	3.2 E-09	6.0 E-08	1.3E-08	
Aerosol *F—type* – AMAD = 1 µm – AMAD = 5 µm	1.0 E-07 1.5 E-07	7.1 E-07 1.0 E-06	1.5 E-07 2.1E-07	1.4 E-09 1.9 E-09	2.8E-08 4.1E-08	5.7E-09 8.2E-09	2.5E-08 3.2E-08
Aerosol *M—type* – AMAD = 1 µm – AMAD = 5 µm	2.2 E-08 2.6 E-08	2.5 E-07 2.4 E-07	2.2 E-08 3.1E-08	1.4E-10 2.0E-10	3.6E-09 5.3E-09	6.5E-10 9.6E-10	4.3E-09 6.4E-09
Aerosol *S—type* – AMAD = 1 µm – AMAD = 5 µm	1.1 E-09 1.5 E-09	3.4 E-08 2.6 E-08	1.1E-09 2.0E-09	1.1E-11 1.8E-11	1.8E-10 3.3E-10	3.8E-11 6.6E-11	3.2E-10 5.0E-10

Inhalation dose coefficients for aerosols also differ according to the aerosol size due to differential deposition in the airways. For workers, an AMAD of 5 μm is commonly assumed while the default value for the public is 1 μm. The ICRP provides dose coefficients according to aerosol sizes in the range of nm to about 10 μm (Fig. 15.2). The dose coefficient variation is about a factor of 3. Assuming a case for which the ^{131}I fraction is supported only by fine aerosols (i.e., typically <0.01 μm), as occurs when gaseous species nucleate or condensate to form nanoparticles, will result in a 2.2 times increase in the inhalation dose from aerosol-bound ^{131}I species compared to the 1 μm aerosol reference size usually considered for members of the public.

In addition, inhalation dose is age-dependent. Table 15.2 gives the dose coefficients for an adult only and Table 15.3 the change in thyroid equivalent dose coefficients with age. The radio-sensitivity of the thyroid gland is especially high for children; among them babies (age <1 year) represent the highest-risk population, while several studies, including that of the survivors of the atomic bombings, have detected no increase in relation to exposures received after the age of 20 years. More detailed information can be found in the ICRP 68 or in its updated version ICRP 137, for workers (occupational intake) and in the ICRP 71 for members of the public (ICRP 68 and ICRP 71 are based on the same bio-kinetic models).

Note that the dose coefficient for ^{129}I is higher than for ^{131}I whatever the species as a result of the five times higher dose coefficient per unit of uptake for ^{129}I than for ^{131}I. A material is assigned to Type F when the deposited materials are readily absorbed into blood from the respiratory tract. Type M is assumed for deposited particulate forms that have intermediate rates of absorption into blood from the respiratory tract, as may be the case of iodine trapped in irradiated fuel fragments.

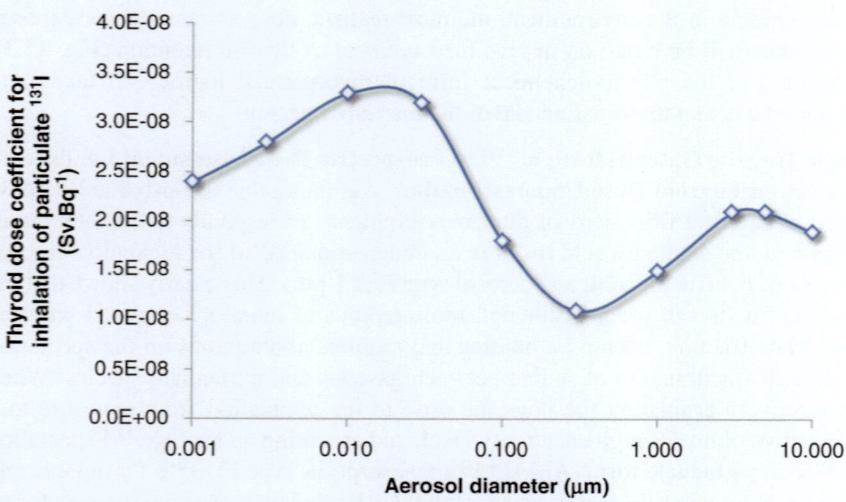

Fig. 15.2 Change in the thyroid equivalent dose coefficients for inhalation of ^{131}I, as a function of the aerosol diameter. After: [29]

Table 15.3 Thyroid equivalent dose coefficients (Sv Bq^{-1}) for inhalation of ^{131}I according to the age at exposure [30]

Iodine species	Age					
	3 months	1 year	5 years	10 years	15 years	Adult
I_2 (gas)	3.3E-6	3.2E-6	1.9E-6	9.5E-7	6.2E-7	3.9E-7
CH_3I (gas)	2.6E-6	2.5E-6	1.5E-6	7.4E-7	4.8E-7	3.1E-7
AMAD 1 μm (aerosol type F)	1.4E-6	1.4E-6	7.3E-7	3.7E-7	2.2E-7	1.5E-7

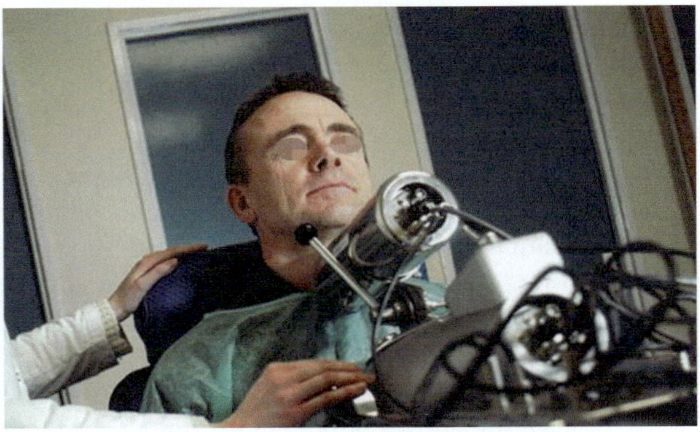

Fig. 15.3 In vivo measurement of ^{131}I activity in the thyroid at IRSN (© IRSN)

A material is assigned to Type S if deposited materials are relatively insoluble in the respiratory tract. In practice, and according to the EU directive 96-29 [31], aerosol-bound iodine isotopes are assigned to absorption type F. After a substantial release of radioiodine in the environment, the most realistic dose assessment for exposed individuals will be based on in vivo measurement of thyroid retention (Fig. 15.3), regardless of the physico-chemical form of incorporated iodine and takes into account the actual thyroidal uptake of the measured person.

Neglecting the Gaseous form of ^{131}I in Prospective Dose Assessment Leads to an Inhalation Thyroid Dose Underestimation Assuming that the airborne ^{131}I activity level deduced from aerosol filter measurements corresponds to the total atmospheric iodine content would result in an underestimation of the inhalation dose by a factor of 6.4 (for an adult and aerosol type F of 1 μm). This clearly shows that the gaseous fraction of iodine, although more tedious to monitor compared with the particulate fraction, cannot be omitted and requires assumptions on the speciation or at least fractionation of iodine between gaseous and particulate species. When prospectively evaluating the potential dose to the population from exposure to a radioactive plume of a given activity level, and assuming an average ^{131}I speciation of 30% in particulate form (AMAD 1 μm, absorption Type F), 35% for organic and 35% for inorganic ^{131}I gaseous species, and ICRP inhalation dose coefficients for an adult will lead to about twice as much higher dose estimate than for the consideration of the particulate form only. In the absence of information regarding the gas/

particle fractionation, it is therefore recommended to assume 50% particulate and 50% gas in the occupational intakes of radionuclides series published by the International Commission on Radiological Protection [28]. Regarding the gaseous speciation, when it is not possible to distinguish inorganic and organic species, the ICRP [28] recommends adoption of the deposition rate for elemental iodine. Based on half-lives and inhalation dose coefficients associated to the various iodine isotopes, those to which attention should be paid first during a nuclear reactor accident are ^{131}I, ^{132}I, and ^{133}I, as well as the precursor of ^{132}I, i.e., ^{132}Te. Indeed, more accurate estimates of radioiodine health impact should also consider decay of tellurium isotopes, especially for inhalation dose assessment in the first few days after a nuclear accident release. Tellurium isotopes present in the releases or in the environment will give the following radioiodine: ^{129}I by filiation of ^{129m}Te ($T_{1/2}$ = 33.6 days) and ^{129}Te ($T_{1/2}$ = 69.6 min), ^{131}I by filiation of ^{131m}Te ($T_{1/2}$ = 30.0 h) and ^{131}Te ($T_{1/2}$ = 25.0 min), ^{132}I by filiation of ^{132}Te ($T_{1/2}$ = 3.2 days; γ_{max} = 228.2 keV with 88.2% abundance), ^{133}I by filiation of ^{133m}Te ($T_{1/2}$ = 55.4 min) and ^{133}Te ($T_{1/2}$ = 12.5 min), and ^{134}I by filiation of ^{134}Te ($T_{1/2}$ = 41.8 min). As for radioiodine with a short half-life of tens of minutes, other tellurium-induced radioiodine isotopes have a too short half-life and can therefore be neglected, since it takes usually several hours after the start of the accident when releases start. Exposures to gas or vapor forms of tellurium are relatively unusual compared with exposures to particulate forms in the environment. From aircraft samplings performed at short distance over the block IV of the Chernobyl NPP, Borisov et al. cited in [32] reported that the gaseous ^{132}Te component varied only 1–8%. It is therefore recommended to consider a particulate form of M-type for tellurium in the absence of specific information [33]. Once inhaled, a part of ^{132}Te ($T_{1/2}$ = 3.2 days) will deposit in the respiratory tract where it will decay to give ^{132}I. The contribution to the thyroid dose from the inhalation of ^{132}Te is substantially higher (by a factor >10) than that from the intake of the same activity of ^{132}I, which is in radioactive equilibrium with ^{132}Te in air [34]. In practice, it means that the inhalation dose contribution from direct intake of ^{132}I can be neglected compared with ^{132}I induced by the intake of ^{132}Te and despite its screening in the respiratory tract. Regarding the Chernobyl accident, Balonov et al. [35] estimated the contribution of ^{132}I to the thyroid doses as being about 30% for person who did not use stable iodine prophylaxis and about 50% for persons who took KI pills on 26–27 April. Regarding the Fukushima accident, Shinkarev et al. [34] demonstrated that on March 12, 2011, the contribution of short-lived radioiodine (^{133}I and ^{132}I induced by the intake and radioactive decay of ^{132}Te) to the thyroid dose from inhalation of the plume might be as great as 30–40% of the dose from ^{131}I, while it could have reached about 10% on March 15, i.e., the day of the main releases. However, the dose from ^{132}I was not considered in the thyroid dose estimation from radioiodine due to the lack of information [36]. In addition to ^{131}I, ^{132}Te, and ^{132}I, other short-lived iodine isotopes (^{123}I, ^{130}I, ^{135}I) can be considered to refine inhalation dose assessment, but this requires prompt measurements. Globally, the contribution of the inhalation exposure pathway to the thyroid dose may be dominant during the first few weeks after the beginning of the releases and as far as the application of dietary consumption restrictions, especially for fresh milk and vegetables, is ensured. No significant health effect is expected for ^{129}I, on account of much lower levels of activity, long radioactive half-life, and rapid biological half-life.

15.4 Management of Nuclear Emergency Situations with Regard to Radioiodine

An effective decision-making process is required in the emergency phase of nuclear events. When a release of radioactive material occurs or is about to happen, prompt protective actions can avoid or mitigate the potential consequences. These actions can be decided and set up preventively in the perspective of a possible future release, or in response to an ongoing release and combine sheltering, administration of stable iodine tablets, evacuation if necessary, and later possible restrictions on foodstuff consumption. In any case, a specific evaluation would be necessary in order to propose protective actions to set up correspondingly to the situation, to the public authorities. For radioactive iodine impact due to the passage of the plume, one has to evaluate the corresponding projected thyroid dose. This protection strategy is driven by comparing those projected doses to the corresponding action guide level (50 mSv as for instance in France, Germany, and many other countries). Transboundary harmonization on reference levels is not yet fully operational. Nevertheless, in Europe, the Herca-Wenra approach proposes that in the very early stages of the crisis, if the analysis of the situation is shared, the bordering countries align themselves with the decisions taken by the country where the accident occurs [37, 38].

From a general point of view, there are several factors that influence the assessment of the radioiodine-induced inhalation dose during an atmospheric release:

- Magnitude of the release, physico-chemical forms of iodine (gaseous inorganic, gaseous organic, aerosol), granulometry of aerosol-bound iodine species;
- Dispersion conditions such as height of the release, meteorological conditions (wind speed, direction, atmospheric stability, boundary layer thickness, presence of rain, fog, etc.), local topography (rural, urban, etc.) and depletion processes (dry deposition, wet scavenging etc.);
- Exposure scenario of individuals such as time of exposure, breathing rate, possible indoor protection factor;
- Biokinetics of the incorporated materials, which depend on factors related to the age and gender (see previous section).

The remaining parameters and the modelings have to be chosen by experts performing dose assessments. Due to major unknowns in number of the parameters and modeling simplifications, especially in the acute phase of the emergency, the projected dose assessment is subject to a significant uncertainty level. In general, a balance is made between model complexity and timely response. Emergency response employs substantial simplifying assumptions that make the evaluation of the consequences become tractable and available rapidly:

- Generally, a Gaussian dispersion modeling is used, with constant and uniform deposition velocities, sometimes with uniform wind speed and direction; gaseous and particulate iodines are generally associated with different dry or wet deposition velocities;
- Representative persons are considered: it consist of a virtual individual, with constant breathing rate, with no benefit of any protection factor during the whole time period of exposure. This fictitious individual is sometimes supposed to remain located at different distances downwind on the plume centerline, where the air-activity concentrations are the highest. For reactor accidents, a 1–2 years old child is the ICRP age category for which the calculated dose is the highest.

Emergency planning ensures the availability of potassium stable iodine (KI) tablets by a pre-distribution performed in the near to medium fields around a nuclear power plant. In France, this distance is set to 20 km. Their administration must be organized as soon as possible in the case of a nuclear accident involving significant radioiodine releases to the atmosphere, to saturate the thyroid and protect it from exposure to radioiodine [39]. Administration is recommended when the thyroid equivalent dose is expected to be higher than 50 mSv. This procedure is complementary to others protective actions such as population sheltering or evacuation. The efficiency of the protection is at its highest when KI tablets are administrated at the time of radioiodine intake and therefore distributed at the start of the accident releases or even better by anticipation; ideally 2 h before inhalation of contaminated air (Table 15.4). ICRP recommends a biological half-life of iodine of 80 days, however a more recent empirical study suggested a biological half-life as short as 66 ± 6 days [41]. From Eq. 15.1, an effective half-life of iodine in the human body (for adults with intact thyroid function) of about 7 days can be deduced (T_{eff} being the effective half-life, T_{bio} the biological half-life, and T_{phys} the physical half-life).

Table 15.4 KI prophylaxis efficiency for different periods between ingestion of KI tablets and intake of radioiodine, after: [40]

Period between administration of KI and intake of radioiodine	Thyroid cover (%)	Factor of exposure reduction
4 days before	5	1.1
3 days before	32	1.5
2 days before	75	4
1 day before	93	14
0 day	97	33
2 h after	80 (65∗)	5
8 h after	40 (15∗)	1.7
16 h after	17	1.2
1 day after	7	1.1

Values are for an area with no iodine deficiency (i.e., normal iodine dietary supply). Values followed by ∗ are for areas with iodine deficiency

$$\frac{1}{T_{\text{eff}}} = \frac{1}{T_{\text{bio}}} + \frac{1}{T_{\text{phys}}} \tag{15.1}$$

However, the efficiency of the KI tablets is about 1 day after ingestion (Fig. 15.4). The following populations are targeted for priority KI tablet administration since they are at highest risk for negative health effects to the thyroid from radioiodine: newborns (< 1 month), infants (1 month–3 years), children (3–12 years), adolescents (12–18 years), pregnant and breastfeeding women [40]. Considering KI tablets of 65 mg (equivalent to 50 mg of I), the dosages are as follows: newborns: ¼ of a KI tablet, dissolved in a liquid; infants: ½ of a KI tablet dissolved in a liquid; children: one tablet; and from 12 years including pregnant women up to 40 years: two KI tablets. In the event of prolonged or repeated exposure by inhalation, public health authorities may advise taking KI tablets more than once. Under such circumstances, neonates (<1 month) and pregnant or breastfeeding women should not be given repeated doses of KI and other protective actions should be considered such as early evacuation, for these particular groups [42]. The administration of iodine would concern in all cases only the emergency and short-term phases. Indeed, if the iodine levels are still too high beyond a few weeks, it also means that the area is very contaminated and that for reasons of external dose, evacuation is advised.

Beside sanitary preoccupations, the management of nuclear emergency situations integrates the communication to mass media. The omission of the gaseous component could be publicly perceived as a deliberate intention to minimize the situation as far as the gaseous predominance will be reminded, and even if levels of activities are of no concern for public health. A mistrust feeling may spread leading to undermine confidence in governmental and expert institutions [43].

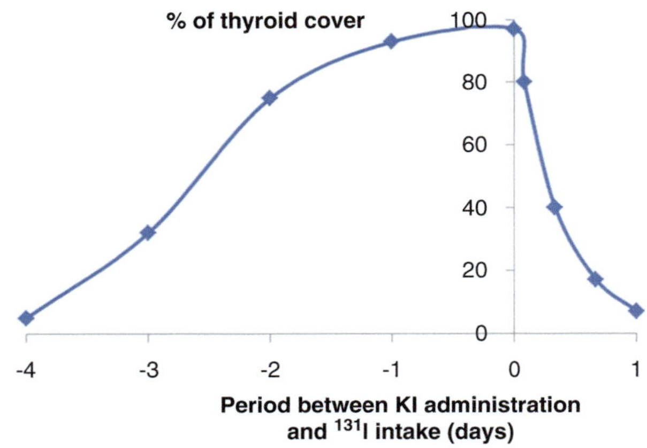

Fig. 15.4 Efficacy of KI as a function of the period between administration and ^{131}I intake. After: [40]

15.5 Anthropogenic Releases and Discharges

Anthropogenic releases and routine discharges of radioiodine in the environment can be categorized in five main sources, namely (1) nuclear weapon tests, (2) nuclear reactor accidents, (3) routine discharges from radiopharmaceutical production[1] and use in nuclear medicine, (4) routine discharges[1] from the nuclear fuel reprocessing industry, and (5) routine discharges[1] from the nuclear power production industry.

15.5.1 Nuclear Weapon Tests (NWT)

First large-scale radioiodine emissions occurred during the nuclear weapons testing era (1945–1980). In total, it is estimated that $0.28–0.98 \cdot 10^{12}$ Bq (or TBq) of ^{129}I and about 6.7–7.8 and even $12.2 \cdot 10^5$ PBq [44] (1 PBq = 10^{15} Bq) of ^{131}I were released by ca. 530–543 NWT (depending on accounting methods) performed in the atmosphere [45, 46] and at ground-level [47, 48]. An approximate rate of 0.17 and 0.28 g of ^{129}I per kiloton TNT equivalent is produced by a nuclear detonation [2], from fission of ^{235}U and ^{239}Pu, respectively. The total yield of atmospheric nuclear weapons tested is about 440 megatons (Mt) among which ca. 182 Mt. came from fission [48]. Globally, the fraction of radionuclides injected by atmospheric detonations into the troposphere quotes about 5% only, the rest into the stratosphere where the residence time of aerosol-bound radionuclides has been estimated 1–2 years. This duration is long enough to allow for a complete decay of ^{131}I before reaching lower tropospheric layers. However, the remaining 5% in the troposphere is of significance for the deposition of short-lived radionuclides such as ^{131}I. This percentage is also closely linked to the bomb yield and latitude because the tropopause height varies with latitude. In the case of a 1 Mt detonation in equatorial region, the percentage of radionuclides released in the troposphere will rise to 65%. Iodine-131 was the most important isotope in fallout with regard to health impact, as demonstrated for instance in the surroundings of the Nevada Test Site (NTS) with an excess risk in thyroid cancers in the US [49]. However, on a global scale, the contribution of ^{131}I to the total effective dose equivalent commitment to the world's population from atmospheric nuclear testing has been estimated in the range 1.4–3.8%, far below the contribution of ^{14}C (70%) [47, 50]. Among radioiodine, only ^{129}I from this period is still present in the environment.

15.5.2 Nuclear Reactor Accidents

In the case of a nuclear reactor accident, radioactive iodine isotopes account for the second largest fraction of emissions, second only to xenon isotopes. In addition, radioiodines belong to the most hazardous class of radionuclides that are released,

[1] Incident releases can be integrated in these categories.

and are responsible for 80–90% of the dose received in the first hours after the start of the accident, primarily due to inhalation of ^{131}I contaminated air (i.e., during the initial emergency situation, and before taking into consideration ingestion of contaminated food). While the dose contribution through contaminated food was significant after Chernobyl, it has been suggested that inhalation was more significant than ingestion in the Fukushima case [51]. Radioiodine and radiotellurium isotopes (see previous section) to be considered in such scenarios have half-lives typically ranging between hours and about 10 days (see Table 15.1). For a loss-of-coolant accident (LOCA), the radioiodine released, as ranked by decreasing order, are: ^{131}I, ^{133}I, ^{135}I, ^{132}I, and to a lesser extent ^{134}I, ^{125}I, ^{129}I, and ^{130}I. The speciation of the iodine species present in the containment (before release in the environment) may differ depending on the accident conditions. Gaseous iodine can be released and aerosols iodine can be also produced by reaction between gaseous iodine and other species such as fission products (Cs, Cd, Ag etc.) or ozone or other air radiolysis products [52]. Various techniques (filtration, adsorption) are deployed including wet purification step to retain the different iodine-containing species, but gaseous CH_3I, unlike I_2, is not significantly absorbed by a liquid phase and requires dedicated filtration steps.

At Windscale Works, Sellafield (U.K.), the graphite-moderated reactor (pile 1) using a gas coolant was operating for the production of plutonium and other materials for the UK weapon program, when it experienced a graphite fire in 1957. The fire caused the release of 0.74–1.8 PBq ^{131}I in the environment (Table 15.5). The damaged (1979) unit 2 of the Three Mile Island (TMI, USA) NPP was a pressurized water reactor (PWR). Despite melting of the fuel, the reactor vessel maintained its integrity, and the damaged fuel was retained inside [69] thus limiting the release into the environment. The unit 4 at the Chernobyl NPP was a graphite-moderated power reactor with pressure channels (RBMK) and the Fukushima Daiichi reactors (units 1, 2, 3, and 4) were BWRs. The Chernobyl and Fukushima accidents were associated with significant core meltdowns, steam or hydrogen explosions, and partial or total loss of the confinement. They were responsible for the highest amounts of radioactive iodine released to the environment (Table 15.5). Estimates of ^{129}I releases during Windscale and TMI accidents are difficult since the ^{129}I generated signal was blurred or encompassed in the global fallout signal [70]. In the case of the Windscale accident, the available activity of ^{129}I was only about $1.5 \cdot 10^{-8}$ times that of ^{131}I. Health consequences were undoubtedly established for the young population exposed to ^{131}I after the Chernobyl accident. The main radionuclides investigated in Europe were ^{99}Mo, ^{103}Ru, ^{106}Ru, ^{110m}Ag, ^{125}Sb, ^{129m}Te, ^{132}Te, ^{131}I, ^{134}Cs, ^{137}Cs, ^{140}Ba, ^{141}Ce, and ^{144}Ce. Other radionuclides (^{60}Co, ^{63}Ni, ^{90}Sr, ^{95}Nb, ^{95}Zr, ^{154,155}Eu, ^{238,239,240,241}Pu, ^{241}Am, and ^{242,243,244}Cm) were also observed in Poland for instance (see Chaps. 1 and 2).

The major released radionuclides from the Fukushima accident were ^{132}Te ($T_{1/2}$ = 3.2 days), ^{131}Te ($T_{1/2}$ = 24.8 min) and their respective progeny ^{132}I and ^{131}I, as well as ^{134}Cs ($T_{1/2}$ = 2.06 years) and ^{137}Cs ($T_{1/2}$ = 30 years). In its source term assessment, the French Institute for Radiological Protection and Nuclear Safety (IRSN) retains the following decreasing importance of released radioiodine (in Bq): ^{131}I,

Table 15.5 Radioiodine amount released during major nuclear accidents

Location, start date, and INES level	^{131}I Activity range and equivalent mass (g)	Others iodine isotopes	% of the core inventory	Ref.
Kyshtym 29/09/1957 (6)	1–10 PBq (0.2–2.0 g)	Unknown	N/A	[53, 54]
Windscale 10/10/1957 (5)	0.74–1.8 PBq (re-evaluation) (0.15–0.38 g)	Unknown	12–15	[55, 56]
Three Mile Island 28/03/1979 (5)	0.48–0.63 TBq (0.10–0.13 g)		$5 \cdot 10^{-5}$	[57, 58]
Chernobyl 26/04/1986 (7)	1760 PBq (380 g)	^{132}I: 1150 PBq from ^{132}Te ^{133}I: 910 PBq ^{134}I: 25 PBq ^{135}I: 250 PBq ^{129}I: 8.5–40 GBq	20–57	[59]
Fukushima 11/03/2011 (7)	90–500 PBq but most often 100–200 PBq (21–42 g)	^{132}I: 7–60 GBq ^{133}I: 42 PBq ^{129}I: 8 GBq (7.0–11.6) ^{132}Te: 60 PBq		[2, 60–68]

^{132}I, ^{133}I, ^{135}I, ^{130}I, ^{134}I, ^{132m}I, ^{129}I, and ^{128}I, for a total amount of 182 PBq. After the Fukushima accident, a cocktail of radionuclides was detectable in Japan among them ^{129m}Te, ^{132}Te, ^{129}I, ^{131}I, ^{132}I, ^{133}I, ^{134}Cs, ^{136}Cs, ^{137}Cs (see [71] for a more complete list). Cesium isotopes (134 and 137), particulate and gaseous ^{131}I were detectable worldwide. At a limited number of European locations equipped with high-volume aerosol samplers and low-level gamma-ray spectrometry, it was also possible to find in addition ^{132}Te and ^{132}I [72], ^{129}Te, ^{129m}Te, and ultra-traces of ^{136}Cs [73] and ^{140}La [74–76].

The immediate aftermath of the Fukushima accident has not yet been fully clarified because no systematic data on exposures of the general public to short-lived radionuclides, especially ^{131}I, were available soon after the releases [36]. Indeed, no fission products could be monitored during 12 days because of blackout of power supply; only dose rates were measured [77]. The lack of information on ambient ^{131}I concentrations in the early stage of an accident can later be partly filled, step by step, but requires strong assumptions on the relationship between ^{131}I and other radionuclides having longer half-lives and on their deposition conditions, once deposited on soil or integrated into the biomass. Among them, ^{129}I is the most interesting since, as an iodine isotope, it is assumed to behave similarly [78] to ^{131}I and its long half-life makes it possible to investigate retrospective evaluation of ^{131}I deposition density and ^{131}I external exposure, long after the accident [79, 80] and even if the uncertainty in the ^{129}I/^{131}I ratio may reach 20% [81] to 30% [82]. Nevertheless, it has to be considered that the ^{129}I background may not be equal to zero when subtracting the pre-accident concentrations, because of ^{129}I from global fallout and ^{129}I fallout from a reprocessing plant (RP), at local or regional scales. The inverse reconstruction of the

^{131}I airborne activities and inhalation dose to the thyroid from ^{129}I deposition may thus be highly uncertain because of additional assumptions on deposition mechanisms. Finally, taking advantage of the affinity of the thyroid gland with iodine, the internal dose from radioiodine, whether inhaled or ingested, is generally evaluated subsequently by in vivo counting methods [83] (thyroid monitoring [84] or whole-body counting) but the drawback is that no distinction between intake from inhalation or from ingestion can be done without additional hypothesis. The knowledge of the iodine concentrations in the air if not in real-time at least within a short delay after sampling is therefore a strong tool for emergency situation management.

15.5.3 Routine Discharges

15.5.3.1 Radiopharmaceutical Production

Medical isotope production facilities (MIPFs) have the highest ^{131}I release authorizations (up to 780 and even 1600 GBq years^{-1} for the two most important ones located in Europe). The most widely produced radioiodine is ^{125}I for radio-immuno-analysis, therapy and diagnostic; and ^{131}I for radiotherapy and diagnosis applications. Medical isotopes facilities also produce ^{133}I in large quantities, ^{123}I and ^{124}I for medical imaging and diagnosis. Up-to-date and detailed information on iodine isotopes for medical purpose can be found in [17]. To cope with an increasing demand of radiopharmaceuticals, the production has raised and is expected to grow [85, 86]. Consequently and despite various abatement techniques [85], the number of ^{131}I detections in the atmosphere close to production units is also increasing [87]. The January–February 2017 European-scale ^{131}I detection event ensued from a combination of routine releases from the main European MIPFs and poor atmospheric dispersion conditions [88].

15.5.3.2 Nuclear Medicine Hospitals

Iodine-131 is a frequently used radionuclide in nuclear medicine with therapeutic (rather than diagnostic) applications. The activity of ^{131}I involved in diagnosis ranges 0.19–3.7 MBq, while it can reach 1.8–9.2 GBq per treatment for a thyroid cancer. A standard amount of ^{123}I administered for diagnosis is 110–220 MBq and will result in an effective dose of 0.5–2.2 mSv, which is close to the average annual natural exposure [17]. One day after the administration of a ^{131}I capsule (NaI) for therapeutic purpose, a patient still exhales ^{131}I, mainly (94% to almost 100%) in organically bound form [89, 90]. Depending on the diagnostic or therapeutic purpose of the treatment, a patient will stay 1–3 days at the hospital before leaving. The first case corresponds to an ambulatory stay for which about 70% of the administrated ^{131}I activity will be excreted in the municipal sewerage system once the patient returns home and taking into account the radioactive decay. In the second case (therapeutic), the percentage of iodine entering the sewerage system is usually insignificant if the nuclear-medicine hospital is (as it is usually the case) equipped with septic tanks in which urines from lavatories are distinctly collected and stored for ^{131}I decay up to 3 months before discharge to municipal sewerage. Regarding excretion, more than 90% of iodine loss from the body is due to renal clearance of iodide [28]. However, part of urines may be directly released in the hospital showers and thus may escape without significant decay before leaving the hospital. Excretion through feces is about 10–12% that of urinary excretion, but feces are usually drained out directly without storage and significant decay as for urines because of the nosocomial hazard that has to be avoided within the hospital premises [88]. Depending on the treatment, almost all the administrated iodine will finally be evacuated from the body within at least a week with an effective half-life of a dozen hours for a carcinoma thyroid, to more than 100 h for patients suffering from hyperthyroidism [91, 92].

15.5.3.3 Sewage, Waste Water Treatment Plants, and Sludge Incinerators

The patient-to-sewage pathway either from hospitals or from home may thus represents a widespread source of ^{131}I to the environment at many locations. Finally, taking account of the activity of urines that escape to the decay tank and that of feces excretion, an estimate of about 20% can be retained for the activity that will be released in the municipal sewer system for patients treated for therapeutic purposes. In addition, excreta may be exempt from regulations that address disposal of radioactivity into municipal sewerage. Wastewater treatment plants (WWTP) settled in watershed area equipped with nuclear-medicine hospitals will thus receive a part of ^{131}I excreted by patients through hospital effluent or from domestic effluents, once returned at home. Iodine-131 entering the WWTP will be covalently bound to organic matter, a reaction that is essentially irreversible [93]. According to Kitto et al. [94], water treatment process and chemical forms (e.g., organic or inorganic) of ^{131}I in the waste influence the ability of the isotope to concentrate in the sludge. The authors confirm that most ^{131}I emission to the atmosphere from a WWTP came originally from excreta entering the plant, then concentrated in dried sludge and finally burned in the WWTP incinerator. According to Hormann and Fischer [95], 20% at most of the ^{131}I inflow activity is retained in the sludge. However, the real amount of remaining ^{131}I greatly depends on the internal structure of the WWTP and time spent by the sludge in the circulation processes prior to their incineration or further treatments, which make it possible for radioactive decay [96]. Other syntheses have reported a ^{131}I retention in sludge ranging 2% to about 20% [97, 98]. Incineration of dried sewage sludge inside WWTP is increasingly applied and encouraged to reduce their transportation cost outside from the plant, to increase the yield of sludge driers or for heating of the premises. Despite the physical decrease of ^{131}I during its residence time inside the plant plus filtration and scrubbing systems to clean the fumes before discharge, WWTP incinerators are prospective ^{131}I re-emission source to the atmosphere. Typical atmospheric ^{131}I emission is about 1%

of the amount entering the plant as wastewater [99] and typical airborne ^{131}I activity of several mBq m^{-3} at 300 m from a WWTP incinerator were observed [100]. Typical daily emissions of gaseous ^{131}I have been estimated between 15 and 60 kBq [94] from such facilities but are of no concern for public health. They are worth mentioning however, since they can explain the detection of ^{131}I in the atmosphere close to medical facilities/hospitals or WWTPs.

15.5.3.4 Reprocessing Plants (RPs)

Nuclear fission of ^{235}U and ^{239}Pu produces ^{129}I and ^{131}I among other radionuclides. At the end of a 4-year activity, the core of a PWR 1300 MW reactor contains about 400 PBq of ^{131}I and 75 GBq of ^{129}I. According to the time spent between de-fueling and reprocessing (about 8–10 years), all ^{131}I has totally disappeared by radioactive decay, whereas ^{129}I is almost entirely present during reprocessing operation. This iodine isotope remains stored in the fuel pellets until the dissolution and solvent extraction steps involved in the reprocessing of spent nuclear fuel which leads to the release of significant amounts of ^{129}I. Despite trapping and filtration of the fumes, a small proportion (ca. 1%) of ^{129}I is emitted to the atmosphere. Main RPs currently in operation are located in La Hague (France) since 1966, in Mayak (Russian Federation) since 1948, and in Rokkasho (Japan) [101] since 2000. Other RPs have ceased operation, such as in the USA: West Valley since 1980, Hanford (1944–1988), Savannah River (1954–1989); in Europe: Sellafield (UK, 1951–ended November 2018 for the Thermal Oxide Reprocessing Plant—THORP), Marcoule (France, 1959–1997), Karlsruhe (Germany, 1970–1990); Tokai-mura (Japan, 1977–2006); Seversk and Zheleznogorsk (Russian Federation, 1956 and 1964–1995). Other plants are likely to be still in operation at the Mayak industrial complex, and in India (Trombay, Tarapur, and Kalpakkam). Detailed estimated amounts of ^{129}I released to the atmosphere are summarized in [102] and in [103]. Due to its long half-life, ^{129}I accumulates in the environment including atmosphere, hydrosphere, and biosphere and will continue to accumulate. The estimate of the naturally produced ^{129}I amount is ca. 250 kg, while the input from anthropogenic activities can be evaluated to more than 6500 kg [2], among which 99% originates from reprocessing activities [104]. It has been assumed by [102] that the average atmospheric ^{129}I releases from the Savannah River facilities was at 58% in a gaseous form. By 2012, the reprocessing plant in La Hague had discharged 4677 kg ^{129}I into the English Channel and 77 kg (~2%) of ^{129}I into the atmosphere, in addition to the 1634 kg ^{129}I into the Irish Sea and 197 kg (~12%) into the atmosphere from the Sellafield RP. Yearly ^{129}I atmospheric release authorizations for La Hague and Sellafield RPs are 18 GBq years^{-1} and 70 GBq years^{-1}, respectively. Actual releases represent about 30–34% for la Hague [105] and 14% for Sellafield, of those authorizations. In addition, the ^{131}I release of Sellafield was about 1.2% of the yearly authorization (37 GBq years^{-1}) [106]. Despite of a higher amount of ^{129}I released by the Fukushima accident ($^{129}I/^{131}I$ ratio isotopic ratio = 31.6 ± 8.9, as of March 15, 2011) [82], it was impossible to discern the Fukushima-derived ^{129}I, while ^{131}I was easily detectable in

Europe after the arrival of the contaminated air masses [107]. This was due to the ^{129}I background originating from European RP discharges [108, 109]. Today most ^{129}I, if not directly deposited on or discharged to the ocean, has migrated in the marine environment. At a more or less local scale, ^{129}I can also be found in the terrestrial environment close to reprocessing facilities or waste disposal repositories. Based on stable iodine distribution in soils, it is apparent that soil-iodine enrichment is limited to about 100–120 km from coastal areas. It can be hypothesized that the same impact can be found for ^{129}I [110].

15.5.3.5 Nuclear Power Plants (NPPs)

In comparison with fuel reprocessing, the nuclear energy industry has much limited authorizations and much lower actual radioiodine discharges to the atmosphere since most iodine remains basically trapped in the fuel rods. Despite the various techniques used to purify the fumes, residual radioiodine activities are released to the atmosphere (about 15 iodine isotopes but mainly ^{131}I and ^{133}I). However, no significant increase in the ^{129}I by AMS measurement technique and no detection of ^{131}I activity in the surroundings of NPPs have been reported [111] to our knowledge, in routine operations. Nedveckaite et al. [112] investigated the reason for thyroid disorders around the Ignalina nuclear power plant and concluded that thyroid disorders ensued from a stable iodine supply deficiency. Taking the case of the French NPPs (pressurized water reactor technology) as an example, the yearly authorized release for radioiodine is 0.4 GBq years^{-1} for each reactor. The average actual discharge for each reactor is estimated to be ca. 0.016 GBq years^{-1}. This estimate is based on accountancy rules established in France according to which the detection limits are considered as being reached even when the measurement result remains below. The actual discharges are likely to be smaller. Additionally, it is supposed that most ^{131}I (CH_3I) discharges occur during gaseous trap efficiency tests.

The contribution of routine emissions to the atmosphere from these different industries or activities (MIPF, NPP, sewage sludge incinerators, etc.) has recently been investigated on the occasion of the early 2017 European-scale detection event of airborne ^{131}I [88]. The expected contributions of various typical routine releases to the ambient atmospheric concentration during this detection event were ranked as follows: MIPF > sewage sludge incinerators > NPP > spontaneous fission of uranium in soil.

Between routine discharges and accident releases, there have been several ^{131}I release incidents within or close to authorization limits, in the recent years from nuclear reactors or MIPFs. Borisov et al. [113] reported the use of combined filter and charcoal packs for the complex analysis of air nearby the St Petersburg NPP (probably in March 1992) with 60–70% of total ^{131}I in gaseous form associated with contamination level of 10^4 Bq m^{-3} [113]. On August 22, 2008, at the Institut des RadioEléments (IRE) in Fleurus (Belgium), 47 GBq of elemental iodine were released at once [114], which corresponded to the yearly authorization limit. This incident was rated 3 on the INES scale (International Nuclear and radiological

Event Scale), but no airborne ^{131}I measurement has been reported. In the time interval between September 8 and November 16, 2011, 342 GBq of ^{131}I, compared to the 1600 GBq yearly authorized, were released from an activation facility at the Institute of Isotopes Ltd., Hungary [115]. This incident was rated 1 on the INES. It led to detections in the range of a few to several tens of μBq m^{-3} in Europe, and even in animal thyroids [84]. Other ^{131}I detections in the μBq m^{-3} to tens μBq m^{-3} range occur from time to time as, for example, in April 2010 in Sweden and Germany; in March 2011 (the week prior to the F1NPP accident) in several countries in Europe; in September to October 2011 as previously indicated; in February 2012 in several European countries; in December 2013 in Scandinavia; in March and May 2015 in Sweden, Finland, Norway, Poland, France, Russian Federation; in January–February 2017 and in January–February 2018, in Europe. Figure 15.5 shows the contribution of various ^{131}I source emissions. Iodine-131 induced by the spontaneous fission of uranium has been added for comparison. Its contribution to ambient airborne activity is much lower than any other anthropogenic emissions [88] and is out of reach of conventional HPGe gamma spectrometry detectors. At distance from the various emission points, only large-scale releases (on the order of magnitude of several tens GBq) can be detectable by low-level gamma spectrometry combined with high-volume aerosol sampling. However, close to routine discharge points (MIPFs, nuclear-medicine hospitals, sewage incinerators), ^{131}I may be detected occasionally.

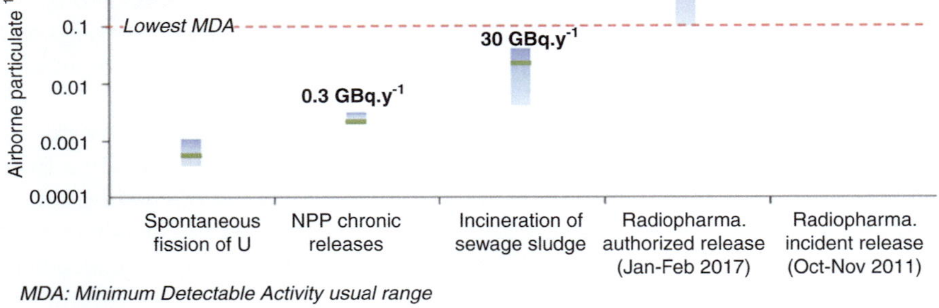

Fig. 15.5 Source apportionment of particulate ^{131}I in the air over Europe in January/February 2017 based on source term estimates. Taken from [88]. Reprinted with permission from the American Chemical Society, © 2018

15.6 Conclusion

The presence of radioiodine in the environment is of major concern for radioprotection issues for workers and members of the public. Regardless of the iodine isotope (^{129}I, ^{131}I, ^{132}I); of the situation (routine release, incident, or accident); and the nuclear activity (radiopharmaceutical production and use, nuclear energy production, nuclear reprocessing), an overall dominant iodine gaseous fraction has to be considered in the global atmosphere at distance from the emission source. The dominancy of the radioiodine gaseous fraction is about 4–5 times higher compared to the aerosol-bound fraction. This pattern applies for Three Mile Island, Chernobyl, and Fukushima Daiichi NPP accidents at more or less long distances from the release points, i.e., where the concentrations of the different radioactive iodine-containing species have significantly decreased because of (1) plume dispersion, (2) deposition mechanisms (dry + wet) involved along the route of the contaminated air masses and according to different deposition velocities linked to the different iodine-containing species, (3) photochemical reaction of gaseous iodine with ozone and NOx leading to oxide particles [8, 116], and (4) the sorption kinetics of gaseous iodine onto ambient aerosols, to a lesser extent.

Exemptions to this scheme (i.e., a dominant radioiodine particulate fraction) may temporarily occur when a high amount of aerosol is produced as during explosions and fires affecting a nuclear reactor or its outer superstructures. At short distances, the release features and/or accident scenario therefore control the distribution of radioiodine, and it is difficult to predict from where and when it will return in favor of the gaseous fraction. This is all more important than at such distances the airborne radioiodine activity levels are likely to be a matter of health issues, and the resulting inhalation dose assessment will be sensitive to the detailed speciation of iodine between gaseous inorganic and organic species, plus aerosol-bound species. During explosions and fires, which is typically what happened during the Chernobyl and Fukushima accident, the amount of particles is likely to enhance the proportion of aerosol-bound iodine. This is also typically what was observed during atmospheric nuclear weapon detonations. In the Fukushima-Daiichi NPP case, explosion phases alternated with venting, scrubbing, and leaking phases, which exhibited a higher gaseous proportion. Methyl iodide (CH_3I) and elemental iodine (I_2) have higher dose coefficients than aerosol-bound species since they settle at 70% and 100%, respectively. At a more detailed stage, is thus of primary importance to know or assess the gaseous chemical speciation (organic and inorganic species) to derive a realistic dose assessment in the early stage of an accident. Indeed, if the gaseous fraction remains unmonitored or ignored, the inhalation dose can be underestimated by a factor of about 6. Recent accident or incident situations involving ^{131}I releases show the need to shift to a higher sampling frequency for most accurate source term estimates and in the case of an undeclared or a priori unknown release, to locate the source location with less uncertainty.

Among the gaseous species, organic gaseous iodine (CH_3I as a representative) is the dominant ^{131}I species in the atmosphere after an accident situation while it is expected to remain lower than elemental iodine (I_2) during a normal operation. Differences in I_2 and HOI contributions were noticed depending on the type of accident (TMI and Chernobyl). In addition to the contribution of iodine-containing species to the inhalation dose assessment, it is important to consider the contribution of ^{132}I induced by decay of ^{132}Te which could lead to a 30% underestimation in the inhalation thyroid dose. Apart from the three main iodine chemical species for which inhalation dose coefficients have been established, little is known about the ^{131}I in the form of hypoiodous acid (HOI) in the atmosphere. Moreover, no inhalation dose coefficient exists for this species. There are some indications of a significant contribution (up to 40%) of $HO^{131}I$ to the total ^{131}I content in the atmosphere when dealing with routine discharges and up to 10% during accident releases. Due to the high photo-dissociation rate of I_2, it has even been proposed to replace I_2 by HOI and $IONO_2$ species with regard to inorganic species to be considered [6]. More investigations on routine releases are welcome. By default, the following distribution can be adopted in the absence of detailed information: 25% for particulate iodine species, 25% for gaseous inorganic species, and 50% for gaseous organic species.

Other topics of interest are directly or indirectly linked to the increasing use of radiopharmaceuticals for the purpose of both diagnostics and treatments of malignant tumors, and the increasing number of patients. The routine radiopharmaceutical production exhibits by far the highest routine releases, and some recent incidents in Europe have highlighted this source. At a lesser extent, atmospheric discharges from waste water treatment plants equipped with incinerators and located on watersheds with nuclear-medicine hospitals can be considered as possible source of ^{131}I emission to the atmosphere, despite cleaning techniques applied to the fumes. Preliminary investigations have already been carried out in Germany [95], Poland, and in the USA [94] but remain limited in number. However, their impact in terms of dose to the residents living nearby is also worth to be assessed.

Despite the Chernobyl and Fukushima accidents, there is still a lack of information and sampling capabilities dedicated to the knowledge of iodine in the atmosphere and especially regarding its gaseous component.

References

1. Saiz-Lopez A, Fernandez RP, Ordóñez C, Kinnison DE, Martín JCG, Lamarque JF, et al. Iodine chemistry in the troposphere and its effect on ozone. Atmos Chem Phys. 2014;14:13119–43.
2. Hou X, Hansen V, Aldahan A, Possnert G, Lind OC, Lujaniene G. A review on speciation of iodine-129 in the environmental and biological samples. Anal Chim Acta. 2009;632:181–96.
3. Trincal J. Modélisation du comportement de l'iode dans l'atmosphère. PhD thesis, Université de Lille, France. 2015.
4. Landis JD, Hamm NT, Renshaw CE, Dade WB, Magilligan FJ, Gartner JD. Surficial redistribution of fallout ^{131}iodine in a small temperate catchment. Proc Natl Acad Sci U S A. 2012;109:4064–9.
5. Kulyukhin SA, Kulemin VV, Rumer IA, Konovalova NA. Gas-phase UV photolysis of $CH_3{}^{131}I$. Radiochemistry. 2005;47:296–300.

6. Filistovic V, Nedveckaite T. Photochemical aspects of the behavior of the atmospheric radio-iodine after the Chernobyl accident. J Radioanal Nucl Chem. 1999;242:75–80.
7. Saiz-Lopez A, Plane JMC, Cuevas CA, Mahajan AS, Lamarque JF, Kinnison DE. Nighttime atmospheric chemistry of iodine. Atmos Chem Phys. 2016;16:15593–604.
8. Jenkin ME, Cox RA, Candeland DE. Photochemical aspects of tropospheric iodine behaviour. J Atmos Chem. 1985;2:359–75.
9. Trincal J, Cantrel L, Cousin F, Fevre-Nollet V, Lebegue P. Impact of atmospheric species reactivity on radioactive gaseous iodine transport in severe accident conditions. WIT Trans Ecol Environ. 2015;198:77–86.
10. Chameides WC, Davis DD. Iodine: its possible role in tropospheric chemistry. J Geophys Res Atmos. 1980;85:7383–98.
11. Brookhaven National Laboratory. Interactive chart of nuclides, NuDat 2.7. 2019. https://www.nndc.bnl.gov/nudat2/. Accessed Jan 2019.
12. Chu SYF, Ekström LP, Firestone RB. Table of radioactive isotopes. 2019. http://nucleardata.nuclear.lu.se/toi/nucSearch.asp. Accessed Jan 2019.
13. ICRP 38. Radionuclides transformations, energy and intensity of emissions. ICRP Publication 38; 1983.
14. Laraweb. Library for gamma and alpha emissions. 2019. http://www.nucleide.org/Laraweb/index.php. Accessed Jan 2019.
15. OECD. JEFF-3.3. 2017. http://www.oecd-nea.org/dbdata/jeff/jeff33/index.html. Accessed Jan 2019.
16. Fan Y, Hou X, Zhou W. Progress on ^{129}I analysis and its application in environmental and geological researches. Desalination. 2013;321:32–46.
17. Shirakami Y. Radioactive Iodine. In: Kaiho T, editor. Iodine chemistry and applications. Hoboken: Wiley; 2015. p. 605–23.
18. Haapanen A. Results of monitoring at Olkiluoto in 2008 – environment: Posiva Working Report 2009-45. 2009.
19. UNSCEAR. Sources and effects of ionising radiation (Report to the General Assembly). New York: United Nations; 2000.
20. Veiga LHS, Holmberg E, Anderson H, Pottern L, Sadetzki S, Adams MJ, et al. Thyroid cancer after childhood exposure to external radiation: an updated pooled analysis of 12 studies. Radiat Res. 2016;185:473–84.
21. Ivanov V, Kashcheev V, Chekin S, Maksioutov M, Tumanov K, Menyajlo A, et al. Results of the thyroid cancer epidemiological survey in Russia following the Chernobyl accident. In: Thyroid cancer and nuclear accidents: long-term aftereffects of Chernobyl and Fukushima. Amsterdam: Academic Press; 2017. p. 88–95.
22. Ivanov VK, Kashcheev VV, Chekin SY, Maksioutov MA, Tumanov KA, Vlasov OK, et al. Radiation-epidemiological studies of thyroid cancer incidence in Russia after the Chernobyl accident (estimation of radiation risks, 1991-2008 follow-up period). Radiat Prot Dosim. 2012;151:489–99.
23. Tronko M, Brenner AV, Bogdanova T, Shpak V, Oliynyk V, Cahoon EK, et al. Thyroid neoplasia risk is increased nearly 30 years after the Chernobyl accident. Int J Cancer. 2017;141:1585–8.
24. Kesminiene A, Evrard AS, Ivanov VK, Malakhova IV, Kurtinaitise J, Stengrevics A, et al. Risk of thyroid cancer among Chernobyl liquidators. Radiat Res. 2012;178:425–36.
25. ICRP 71. Age-dependent doses to members of the public from intake of radionuclides. Part 4. Inhalation dose coefficients. ICRP Publication 71; 1995.
26. ICRP 68. Dose coefficients for intakes of radionuclides by workers. ICRP Publication 68; 1994.
27. Morgan A, Morgan DJ, Evans JC, Lister BAJ. Studies on the retention and metabolism of inhaled methyl iodide-II: metabolism of methyl iodide. Health Phys. 1967;13:1067–74.
28. ICRP 137. Occupational intakes of radionuclides. ICRP Publication 137; 2017.

29. ICRP 66. Human respiratory tract model for radiological protection. ICRP Publication 66; 1994.
30. ICRP 72. Age-dependent doses to the members of the public from intake of radionuclides - part 5 compilation of ingestion and inhalation coefficients. ICRP Publication 72; 1996.
31. EUR-lex. Council Directive 96/29/Euratom of 13 May 1996 laying down basic safety standards for the protection of the health of workers and the general public against the dangers arising from ionizing radiation. 1996. https://eur-lex.europa.eu/eli/dir/1996/29/oj. Accessed Jan 2019.
32. Ogorodnikov BI, Budyka AK, Pazukhin ÉM, Krasnov VA. Aerosol emissions from the destroyed power-generating unit of the Chernobyl nuclear power plant in 1986 and 2003-2005. Atomic Energy. 2006;100:264–70.
33. ICRP. ANNEX B: deposition, characterisation, and sampling of radioactive aerosols. Ann ICRP. 2002;32:139–306.
34. Shinkarev SM, Kotenko KV, Granovskaya EO, Yatsenko VN, Imanaka T, Hoshi M. Estimation of the contribution of short-lived radioiodines to the thyroid dose for the public in case of inhalation intake following the Fukushima accident. Radiat Prot Dosim. 2015;164:51–6.
35. Balonov M, Kaidanovsky G, Zvonova I, Kovtun A, Bouville A, Luckyanov N, et al. Contributions of short-lived radioiodines to thyroid doses received by evacuees from the Chernobyl area estimated using early in vivo activity measurements. Radiat Prot Dosim. 2003;105:593–9.
36. Tokonami S, Hosoda M. Thyroid equivalent doses for evacuees and radiological impact from the Fukushima nuclear accident. Radiat Meas. 2018;119:74–9.
37. European Commission. Radiation Protection No. 165; Medical effectiveness of iodine prophylaxis in a nuclear reactor emergency situation and overview of European practices. 2010. https://ec.europa.eu/energy/sites/ener/files/documents/165.pdf.
38. HERCA. New European approach for cross-border emergency preparedness. 2014. http://www.herca.org/herca_news.asp?newsID=41. Accessed Jan 2019.
39. Reiners C, Schneider R. Potassium iodide (KI) to block the thyroid from exposure to I-131: current questions and answers to be discussed. Radiat Environ Biophys. 2013;52:189–93.
40. Ministry of Health and Long-Term Care (Canada). Potassium Iodide (KI) Guidelines. 2014. http://www.health.gov.on.ca/en/pro/programs/emb/rhrp/docs/ki_guidelines.pdf. Accessed Jan 2019.
41. Kramer GH, Hauck BM, Chamberlain MJ. Biological half-life of iodine in adults with intact thyroid function and in athyreotic persons. Radiat Prot Dosim. 2002;102:129–35.
42. WHO. Use of potassium iodide for thyroid protection during nuclear or radiological emergencies. Geneva: WHO; 2011. https://www.who.int/ionizing_radiation/pub_meet/tech_briefings/potassium_iodide/en/.
43. Petrova K, Jankovec M, Fojtíkova I, Hůlka J. New challenges in crisis communication – the results of sociological survey in the Czech Republic. In: Proceedings of the RICOMET 2017 conference: social and ethical aspects of decision-making in radiological risk situations. IAEA: Vienna, Austria. 27–29th June 2017.
44. Izrael YA. Radioactivity in the environment, vol. 3. New York: Elsevier; 2002.
45. Hu Q, Moran JE. Iodine. In: Atwood DA, editor. Radionuclides in the environment. Chichester: Wiley; 2010. p. 165–78.
46. NCRP. 129I: evaluation releases from nuclear power generation. Report NCRP no. 75. Bethesda: National Council on Radiation Protection and Measurements; 1983.
47. Prăvălie R. Nuclear weapons tests and environmental consequences: a global perspective. Ambio. 2014;43:729–44.
48. UNSCEAR. United Nations Scientific Committee on the Effects of Atomic Radiation. Reports to the General Assembly of the United Nations. New York: United Nations; 1982.
49. Gilbert ES, Huang L, Bouville A, Berg CD, Ron E. Thyroid cancer rates and ^{131}I doses from Nevada atmospheric nuclear bomb tests: an update. Radiat Res. 2010;173:659–64.

50. UNSCEAR. Report to the general assembly (Annex B—Exposures from man-made sources of radiation). 1993. http://www.unscear.org/docs/reports/1993/1993c_pages%2091-120.pdf. Accessed Jan 2019.
51. Steinhauser G, Chávez-Ortega M, Vahlbruch J-W. Japanese food data challenge the claimed link between Fukushima's releases and recently observed thyroid cancer increase in Japan. Sci Rep. 2017;7:10722.
52. Colombani J, Gregoire AC, Morin S. Main findings of the IRSN experimental programs performed on iodine chemistry in severe accident conditions. In: Proceedings of the International OECD/NEA-NUGENIA/SARNET Workshop April 1, 2015. Marseille (France). 2015.
53. Nikipelov BV, Romanov GN, Buldakov LA, Babaev NS, Kholina YB, Mikerin EI. Accident in the southern Urals on 29 September 1957. International Atomic Energy Agency Report INFCIRC-368, Vienna; 1989.
54. Wakeford R. A double diamond anniversary - Kyshtym and Windscale: the nuclear accidents of 1957. J Radiol Prot. 2017;37:E7–E13.
55. Garland JA, Wakeford R. Atmospheric emissions from the Windscale accident of October 1957. Atmos Environ. 2007;41:3904–20.
56. McNally RJQ, Wakeford R, James PW, Basta NO, Alston RD, Pearce MS, et al. A geographical study of thyroid cancer incidence in north-west England following the Windscale nuclear reactor fire of 1957. J Radiol Prot. 2016;36:934–52.
57. Levin RJ, De Simone NF, Slotkin JF, Henson BL. Incidence of thyroid cancer surrounding three mile island nuclear facility: the 30-year follow-up. Laryngoscope. 2013;123:2064–71.
58. Rogovin M, Frampton GT, Cornell EK, DeYoung RC, Budnitz R, Norry P. Three Mile Island: a report to the commission and the public. NUREG/CR 1250 volume 1. Washington: US NRC; 1980. 184 pp
59. UNSCEAR. Exposures and effects of the Chernobyl accident (Annex J). New York: United Nations; 2000.
60. IAEA. The Fukushima Daiichi accident. Technical volume 1/5. Description and context of the accident. Vienna: IAEA STI/PUB/1710. 2015.
61. IRSN. Information report, 22 March 2011. 2011. www.irsn.fr/EN/news/Documents/IRSN_fukushima-radioactivityreleased-assessment-EN.pdf. Accessed Jan 2019.
62. Mathieu A, Korsakissok I, Quélo D, Groëll J, Tombette M, Didier D, et al. Atmospheric dispersion and deposition of radionuclides from the Fukushima Daiichi nuclear power plant accident. Elements. 2012;8:195–200.
63. MEXT. Radioactive substance emission data, press release, October 20, 2011. 2011. www.meti.go.jp/press/2011/10/20111020001/20111020001.pdf. (in Japanese). Accessed Jan 2019.
64. NISA. Evaluation of the status of reactor cores in units 1–3 of the Fukushima Daiichi nuclear power plant. 2011. http://www.meti.go.jp/press/2011/06/20110606008/20110606008-2.pdf. (in Japanese). Accessed Jan 2019.
65. Povinec PP, Gera M, Holý K, Hirose K, Lujaniené G, Nakano M, et al. Dispersion of Fukushima radionuclides in the global atmosphere and the ocean. Appl Radiat Isot. 2013;81:383–92.
66. Saunier O, Mathieu A, Didier D, Tombette M, Quélo D, Winiarek V, et al. An inverse modeling method to assess the source term of the Fukushima nuclear power plant accident using gamma dose rate observations. Atmos Chem Phys. 2013;13:11403–21.
67. Steinhauser G, Brandl A, Johnson TE. Comparison of the Chernobyl and Fukushima nuclear accidents: a review of the environmental impacts. Sci Total Environ. 2014;470–471:800–17.
68. UNSCEAR. Sources, effects and risks of ionizing radiation. UNSCEAR 2013 Report, Volume I, Scientific Annex A. Levels and Effects of Radiation Exposure due to the Nuclear Accident after the 2011 Great East-Japan Earthquake and Tsunami New York. 2014.
69. Yamada M. A brief review of environmental impacts and health effects from the accidents at the three Mile Island, Chernobyl and Fukushima Daiichi nuclear power plants. Radiat. Emerg. Med. 2012;1:33–9.

70. Gallagher D, McGee EJ, Mitchell PI, Alfimov V, Aldahan A, Possnert G. Retrospective search for evidence of the 1957 Windscale fire in NE Ireland using ^{129}I and other long-lived nuclides. Environ Sci Technol. 2005;39:2927–35.
71. Steinhauser G. Fukushima's forgotten radionuclides: a review of the understudied radioactive emissions. Environ Sci Technol. 2014;48:4649–63.
72. Lozano RL, Hernandez-Ceballos MA, Adame JA, Casas-Ruiz M, Sorribas M, San Miguel EG, et al. Radioactive impact of Fukushima accident on the Iberian Peninsula: evolution and plume previous pathway. Environ Int. 2011;37:1259–64.
73. Knetsch GJ. Environmental radioactivity in the Netherlands. Results in 2011. RIVM report 610891004/2013. 2013.
74. De Vismes OA, Gurriaran R, Cagnat X, Masson O. Fission product activity ratios measured at trace level over France during the Fukushima accident. J Environ Radioact. 2013;125:6–16.
75. Gudelis A, Druteikiene R, Lujaniene G, Maceika E, Plukis A, Remeikis V. Radionuclides in the ground-level atmosphere in Vilnius, Lithuania, in March 2011, detected by gamma-ray spectrometry. J Environ Radioact. 2012;109:13–8.
76. Lujanienė G, Byčenkienė S, Povinec PP, Gera M. Radionuclides from the Fukushima accident in the air over Lithuania: measurement and modelling approaches. J Environ Radioact. 2012;114:71–80.
77. Doizi D, Reymond la Ruinaz S, Haykal I, Manceron L, Perrin A, Boudon V, et al. Analytical measurements of fission products during a severe nuclear accident. EPJ Web Conf. 2018;170:08005.
78. Xu S, Zhang L, Freeman SPHT, Hou X, Shibata Y, Sanderson D, et al. Speciation of radiocesium and radioiodine in aerosols from Tsukuba after the Fukushima nuclear accident. Environ Sci Technol. 2015;49:1017–24.
79. Muramatsu Y, Matsuzaki H, Toyama C, Ohno T. Analysis of ^{129}I in the soils of Fukushima prefecture: preliminary reconstruction of ^{131}I deposition related to the accident at Fukushima Daiichi nuclear power plant (FDNPP). J Environ Radioact. 2015;139:344–50.
80. Pietrzak-Flis Z, Krajewski P, Radwan I, Muramatsu Y. Retrospective evaluation of ^{131}I deposition density and thyroid dose in Poland after Chernobyl accident. Health Phys. 2003;84:698–708.
81. Yang G, Tazoe H, Yamada M. Can ^{129}I track ^{135}Cs, ^{236}U, ^{239}Pu, and ^{240}Pu apart from ^{131}I in soil samples from Fukushima prefecture, Japan? Sci Rep. 2017;7:15369.
82. Miyake Y, Matsuzaki H, Fujiwara T, Saito T, Yamagata T, Honda M, et al. Isotopic ratio of radioactive iodine ($^{129}I/^{131}I$) released from Fukushima Daiichi NPP accident. Geochem J. 2012;46:327–33.
83. Kim E, Kurihara O, Tani K, Ohmachi Y, Fukutsu K, Sakai K, et al. Intake ratio of ^{131}I to ^{137}CS derived from thyroid and whole-body doses to Fukushima residents. Radiat Prot Dosim. 2015;168:408–18.
84. Steinhauser G, Merz S, Kübber-Heiss A, Katzlberger C. Using animal thyroids as ultra-sensitive biomonitors for environmental radioiodine. Environ Sci Technol. 2012;46:12890–4.
85. Doll CG, Sorensen CM, Bowyer TW, Friese JI, Hayes JC, Hoffmann E, et al. Abatement of xenon and iodine emissions from medical isotope production facilities. J Environ Radioact. 2014;130:33–43.
86. Karpov Institute of Physical Chemistry. Activity report. 2017. http://www.rosatom.ru/upload/iblock/334/334f8a20cd4f2b02cb25e58b48190bd8.pdf. (in Russian). Accessed Jan 2019.
87. Ageeva NV, Kim VM, Vasilieva KI, Katkova MN, Volokitin AA, Polyanskaya ON. Long-term monitoring airborne I-131 in the surface layer in Obninsk city. Radiation & Risk. 2015;24:96–107.
88. Masson O, Steinhauser G, Wershofen H, Mietelski JW, Fischer HW, Pourcelot L, et al. Potential source apportionment and meteorological conditions involved in airborne ^{131}I detections in January/February 2017 in Europe. Environ Sci Technol. 2018;52:8488–500.
89. Schomäcker K, Sudbrock F, Fischer T, Dietlein M, Kobe C, Gaidouk M, et al. Exhalation of ^{131}I after radioiodine therapy: measurements in exhaled air. Eur J Nucl Med Mol Imaging. 2011;38:2165–72.

90. Sudbrock F, Fischer T, Zimmermanns B, Drzezga A, Schomäcker K. Exhalation of ^{131}I after radioiodine therapy: Dosimetric considerations based on measurements in exhaled air. J Environ Radioact. 2017;166:162–5.
91. Ravichandran R, Binukumar JP, Al Saadi A. Estimation of effective half life of clearance of radioactive iodine (^{131}I) in patients treated for hyperthyroidism and carcinoma thyroid. Indian Journal of Nuclear Medicine. 2010;25:49–52.
92. Willegaignon J, Malvestiti LF, Guimarães MI, Sapienza MT, Endo IS, Neto GC, et al. ^{131}I effective half-life (T_{eff}) for patients with thyroid cancer. Health Phys. 2006;91:119–22.
93. Yeager CM, Amachi S, Grandbois R, Kaplan DI, Xu C, Schwehr KA, et al. Microbial transformation of iodine: from radioisotopes to iodine deficiency. Adv Appl Microbiol. 2017;101:83–136.
94. Kitto ME, Fielman EM, Fielman SE, Gillen EA. Airborne ^{131}I at a background monitoring site. J Environ Radioact. 2005;83:129–36.
95. Hormann V, Fischer HW. The physicochemical distribution of ^{131}I in a municipal wastewater treatment plant. Internal report, University of Bremen; 2017. pp. 1–17.
96. Hormann V, Fischer HW. A simple compartment model for the dynamical behavior of medically derived ^{131}I in a municipal wastewater treatment plant. Environ Sci Technol. 2018;52:9235–42.
97. Fenner FD, Martin JE. Behavior of $Na^{131}I$ and Meta (^{131}I) Iodobenzylguanidine (MIBG) in municipal sewerage. Health Phys. 1997;73:333–9.
98. Titley JG, Carey A, Crockett GM, Ham GJ, Harvey MP, Mobbs SF, et al. Investigation of the sources and fate of radioactive discharges to public sewers, environment agency R&D technical report P288. Bristol: Environment Agency; 2000.
99. Masson O, Baeza A, Bieringer J, Brudecki K, Bucci S, Cappai M, et al. Tracking of airborne radionuclides from the damaged Fukushima Dai-ichi nuclear reactors by European networks. Environ Sci Technol. 2011;45:7670–7.
100. Kitto ME, Fielman EM, Hartt GM, Gillen EA, Semkow TM, Parekh PP, et al. Long-term monitoring of radioactivity in surface air and deposition in New York state. Health Phys. 2006;90:31–7.
101. Japan Nuclear Fuel Limited. Reprocessing. 2019. https://www.jnfl.co.jp/en/business/reprocessing/. Accessed Jan 2019.
102. Reithmeier H, Lazarev V, Rühm W, Nolte E. Anthropogenic ^{129}I in the atmosphere: overview over major sources, transport processes and deposition pattern. Sci Total Environ. 2010;408:5052–64.
103. Eslinger PW, Napier BA, Anspaugh LR. Representative doses to members of the public from atmospheric releases of ^{131}I at the Mayak production association facilities from 1948 through 1972. J Environ Radioact. 2014;135:44–53.
104. Xing S, Hou X, Aldahan A, Possnert G, Shi K, Yi P, et al. Iodine-129 in snow and seawater in the Antarctic: level and source. Environ Sci Technol. 2015;49:6691–700.
105. Orano. Rapport d'information sur la sûreté nucléaire et la radioprotection du site AREVA la Hague (in French). 2014.
106. Sellafield Ltd. (2015) Monitoring our Environment. Discharges and environmental monitoring. Annual report.
107. Zhang L, Hou X, Xu S. Speciation of ^{127}I and ^{129}I in atmospheric aerosols at Risø, Denmark: insight into sources of iodine isotopes and their species transformations. Atmos Chem Phys. 2016;16:1971–85.
108. Daraoui A, Riebe B, Walther C, Wershofen H, Schlosser C, Vockenhuber C, et al. Concentrations of iodine isotopes (^{129}I and ^{127}I) and their isotopic ratios in aerosol samples from northern Germany. J Environ Radioact. 2016;154:101–8.
109. Xu S, Freeman SPHT, Hou X, Watanabe A, Yamaguchi K, Zhang L. Iodine isotopes in precipitation: temporal responses to ^{129}I emissions from the Fukushima nuclear accident. Environ Sci Technol. 2013;47:10851–9.
110. Fuge R. Iodine deficiency: an ancient problem in a modern world. Ambio. 2007;36:70–2.

111. Hou XL, Dahlgaard H, Nielsen SP, Kucera J. Level and origin of Iodine-129 in the Baltic Sea. J Environ Radioact. 2002;61:331–43.
112. Nedveckaite T, Motiejunas S, Kucinskas V, Mazeika J, Filistovic V, Jusciene D, et al. Environmental releases of radioactivity and the incidence of thyroid disease at the Ignalina nuclear power plant. Health Phys. 2000;79:666–74.
113. Borisov NB, Budyka AK, Verbov VV, Ogorodnikov BI. Monitoring of radioiodine in air. J Aerosol Sci. 1994;25:271–2.
114. Vandecasteele CM, Sonck M, Degueldre D. Rejet accidentel d'iode-131 par l'IRE sur le site de Fleurus: retour d'expérience de l'autorité de sûreté belge. Radioprotection. 2011;46:159–73. (in French)
115. European Commission. Technical Report HU-12/01. 2012. https://ec.europa.eu/energy/sites/ener/files/documents/tech_report_hungary_2012_en.pdf. Accessed Jan 2012.
116. Garland JA. The adsorption of iodine by atmospheric particles. Journal of Nuclear Energy. 1967;21:687–700.

Chapter 16
Utilization of Radioxenon Monitoring to Aid Severe Nuclear Accident Response

Steven Biegalski

Abstract Radioxenon isotopes are produced with high yield from nuclear fission. The noble gas properties facilitate migration through nuclear fuel. At fuel temperatures present in a severe nuclear accident, the release fraction of xenon isotopes from the UO_2 matrix approaches 100%. Consequently, a breech in fuel cladding integrity during a severe nuclear accident will likely lead to a radioxenon release. This radioxenon release provides an early signature indicating fuel damage. Fuel was significantly damaged at the Windscale (1957), Three Mile Island (1979), Chernobyl (1986), and Fukushima Daiichi (2011) nuclear reactor accidents, and radioxenon releases were early indicators of the extent of fuel damage for these events. Radioxenon isotopes are an optimum indicator as their half-lives are long enough to be detected but not too long to cause significant environmental background accumulation. The radioxenon isotopes are also readily detectable via modern radiation monitoring equipment.

The chapter will describe the properties of radioxenon, identify the release mechanisms of radioxenon from nuclear fuel, summarize the radioxenon releases from historic severe nuclear accidents, and review the ways that radioxenon may be utilized to aid responders in a severe nuclear accident.

Keywords Radioxenon · Noble gas · Nuclear reactor accident · Forensics · Nuclear fuel release

---

S. Biegalski (✉)
Nuclear and Radiological Engineering and Medical Physics, Georgia Institute of Technology, Atlanta, GA, USA
e-mail: steven.biegalski@me.gatech.edu

G. Steinhauser et al. (eds.), *Nuclear Emergencies*,
Current Topics in Environmental Health and Preventive Medicine,
https://doi.org/10.1007/978-981-13-8327-4_16

16.1 Radioxenon

Noble gas fission products are of special interest as an indicator of fuel damage within a severe nuclear accident due to their natural physical property of high mobility within nuclear fuel. Xenon is present naturally in the atmosphere at a concentration of 0.087 ± 0.001 parts per million (μl l^{-1}) [1]. Table 16.1 lists the nine naturally present stable xenon isotopes along with their radiative capture cross-sections. Out of these isotopes, ^{124}Xe, ^{134}Xe, and ^{136}Xe are predicted to undergo double beta decay [2, 3]. The radiative neutron capture cross-sections are of particular note. Stable xenon isotopes including ^{124}Xe, ^{129}Xe, ^{130}Xe, and ^{131}Xe have particularly high probabilities for neutron absorption. As a result, Xe in air and dissolved in water will activate to radioxenon isotopes in the presence of neutrons from fission.

Table 16.2 lists the radioxenons of interest to this work. While Xe isotopes range from ^{110}Xe to ^{147}Xe, the radioxenon isotopes listed in Table 16.2 are expected to be produced within a nuclear reactor environment. They are also produced from other sources, so monitoring for radioxenon should always include a source assessment. While other Xe isotopes may be produced in a reactor, their half-lives are too short for a realistic chance of monitoring them.

Table 16.2 also shows that each of the radioxenon isotopes of interest, gamma rays, X-rays, and electrons. The electron emissions are either in the form of conversion electrons or beta particles. The X-rays listed are the primary $K_{\alpha 1}$ and $K_{\alpha 2.}$ The X-rays are in coincidence with the conversion electrons. As a result, all of the radioxenons of interest produce signals that are in coincidence with the ca. 30-keV X-rays. For radioxenons such as ^{131m}Xe and ^{133m}Xe, the coincidence signal with X-rays is the primary decay path and the main means for detection. As a result, it is of primary importance that all the radioxenon signatures be well understood.

Primary fission products to be monitored from a nuclear fuel leak within a severe nuclear accident are ^{131m}Xe, ^{133m}Xe, ^{133}Xe, and ^{135}Xe. ^{137}Xe has too short of a half-life for realistic monitoring. ^{125}Xe, ^{127}Xe, and ^{129m}Xe have very low fission yields, so the primary path for production of these radioxenons is through activation of ^{124}Xe, ^{126}Xe, and ^{128}Xe, respectively.

Table 16.1 Stable isotopes of xenon[a]

Target isotope	Natural abundance (%)	Maxwellian average at 0.0253 eV	Resonance integral	Fission spectrum average	14 MeV
^{124}Xe	0.10	146.7 b	2968 b	570.9 mb	2.741 mb
^{126}Xe	0.09	3.784 b	23.33 b	307.9 mb	1.797 mb
^{128}Xe	1.92	7.090 b	12.46 b	97.02 mb	1.050 mb
^{129}Xe	26.4	18.63 b	255.3 b	40.16 mb	1.000 mb
^{130}Xe	4.1	23.05 b	17.76 b	100.5 mb	1.053 mb
^{131}Xe	21.2	75.48 b	899.3 b	40.98 mb	1.000 mb
^{132}Xe	26.9	399.0 mb	4.503 b	13.34 mb	1.001 mb
^{134}Xe	10.4	234.9 mb	616.1 mb	9.466 mb	1.001 mb
^{136}Xe	8.9	230.5 mb	141.1 mb	0.5372 mb	1.000 mb

[a]Data compiled from http://atom.kaeri.re.kr/

Table 16.2 Radioxenon isotopes of interest[a]

Isotope	Half-life	Primary decay mechanism	Primary gamma rays (keV)	X-rays (keV) ($K_{\alpha 1}$ and $K_{\alpha 2}$)	Conversion electrons (keV)	Cumulative yield from ^{235}U thermal fission (%)
^{125}Xe	16.9 h	Electron capture	188.4, 243.4	28.612, 28.317	1408.62, 1463.59, 471.11	6.32×10^{-20}
^{127}Xe	36.4 days	Electron capture	202.8, 172.1, 375.0	28.612, 28.317	459.54, 287.41	1.81×10^{-15}
^{129m}Xe	8.88 days	Internal transition	39.58, 196.6	29.782, 29.461	162.00, 191.11	3.47×10^{-10}
^{131m}Xe	11.934 days	Internal transition	163.93	29.782, 29.461	129.369, 158.477	0.0317
^{133m}Xe	2.19 days	Internal transition	233.22	29.782, 29.461	198.660, 227.768	0.194
^{133}Xe	5.243 days	β^-	80.99	30.973, 30.625	45.012, 75.283	6.70
^{135m}Xe	15.29 min	Internal transition	526.56	29.782, 29.461	492.000, 521.108	1.21
^{135}Xe	9.14 h	β^-	249.79, 608.19	30.973, 30.625	213.809	6.53
^{137}Xe	3.818 min	β^-	455.49	30.973, 30.625	419.505	6.11

[a]Data compiled from http://atom.kaeri.re.kr/, http://ie.lbl.gov/toi/nucSearch.asp and http://www.nndc.bnl.gov/nudat2/indx_dec.jsp

Table 16.3 Radionuclide releases from anthropogenic nuclear activities [4]

Source	Typical order of magnitude of radioxenon release
Hospitals	~10^6 Bq days^{-1}
Nuclear power plants	~10^9–10^{11} Bq days^{-1}
Radiopharmaceutical facilities	~10^{11}–10^{13} Bq days^{-1}
1 kton nuclear explosion underground	0–10^{15} Bq event^{-1}
1 kton nuclear explosion atmospheric	~10^{16} Bq event^{-1}

Radioxenon in the atmosphere has both anthropogenic and natural origins. Table 16.3 summarizes radioxenon releases from anthropogenic facilities.

Radioxenon is also present in the environment produced from natural sources. The two primary natural sources of radioxenon include cosmic-ray activation and spontaneous fission [5, 6]. The dominant source of subsurface and oceanic natural radioxenon is the spontaneous fission of ^{238}U. Cosmic-ray activation of xenon is the dominant source term of radioxenon in the atmosphere. Activity concentrations of these natural radiation sources are low and challenging to detect from natural sources [7].

16.2 Fission Product Releases from Nuclear Fuel

In nuclear accidents, fuel temperatures may range from 600 °C up to in exceedance of 2400 °C. The melting temperature of UO_2 fuel is 2850 °C and the melting point of Zircalloy 4 cladding is 1850 °C. While these may seem like hazard points for the degradation of the nuclear fuel, fuel integrity starts to degrade by U–Zr eutectic flows starting as low as 1227 °C [8]. This leads to fuel rod ballooning and potential rupture. Cladding oxidation becomes significant around 1000 °C via the reaction shown in Eq. (16.1).

$$Zr + 2H_2O \rightarrow ZrO_2 + 2H_2 + \text{heat} \tag{16.1}$$

Zirconium oxidation is exothermic, producing 6.5 MJ kg_{Zr}^{-1} and produces hydrogen that may lead to a hydrogen deflagration or detonation. The heat produced from this reaction increases temperatures and further degrades fuel integrity. The zirconium oxidation continues until it is quenched through ample water supply or it becomes reactant limited. Once the structural integrity of the fuel cladding is breached, a pathway is open to the release of fission products to the reactor vessel.

Noble gasses begin a path to the surface of the fuel starting at the time of production. Xenon gases are produced from the fission process as shown in Table 16.2. Xenon atoms are born in the crystal lattice of the UO_2 fuel with significant kinetic energy. The Xe atom transfers its kinetic energy to surrounding atoms as it slows down, creating on the order of 10^4–10^5 atom displacements. While most of the displaced atoms quickly relocate [9], some point defects are created. Eventually intergranular gas bubbles start to form. Xe produced during fission of uranium has low solubility in UO_2. Hence, after a relatively short irradiation period, a large number of fission gas bubbles are generated within the fuel grain [10]. Through diffusion, Xe gas also transports to grain boundaries and forms additional bubbles there. The grain boundaries serve as a system of paths that facilitate noble gas and volatile elements to transport to the fuel surface. At fuel temperatures <1000 °C, diffusion of Xe is athermal, independent of temperature, but is enhanced by the fission process [11].

In severe nuclear reactor accident conditions, the fuel and cladding become defective. These deformations increase the oxygen-to-uranium ratio. This increases the diffusion rate of the fission products in the fuel resulting in a larger fraction of the fission products being released. The VERCOS experiments examined fission product release in severe accident conditions. These tests showed that the fractional release of Xe ranged from 33 to 68% as temperatures ranged from 1900 °C to 2297 °C for experiments with Xe release reported [12].

Figure 16.1 shows Knudsen effusion mass spectrometric measurements of irradiated fuel showing that xenon is released in stages as temperature increases [13]. These data show the fractional release of xenon as a function of temperature from irradiated mixed-oxide fuel with a burn-up of about 5.5 at.%. At temperatures below 1700 °C, the Xe release is dominated by the gas accumulated at the grain boundaries. Above 1700 °C, the Xe trapped in intragranular bubbles is released. The release fraction of Xe gas is close to 1.0 by 2800 °C.

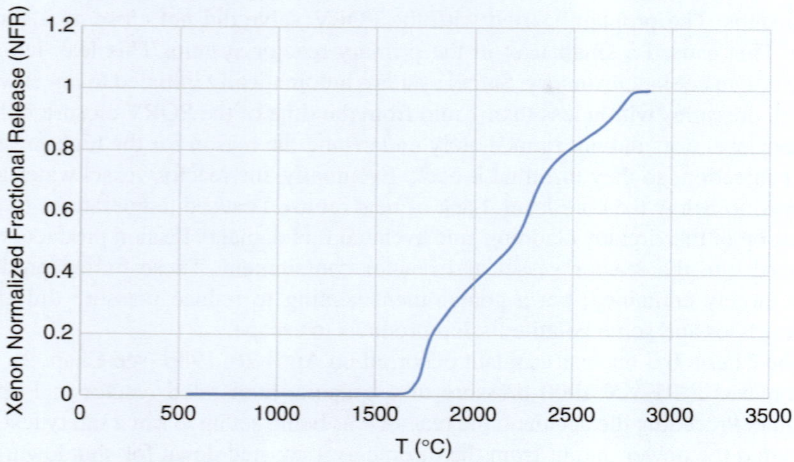

Fig. 16.1 Xenon normalized fractional release (NFR)

16.3 Nuclear Accidents

Radioxenon gas release has been measured from all the main nuclear accidents to date. These accidents include Windscale (1957), Three Mile Island (1979), Chernobyl (1986), and Fukushima Daiichi (2011). Detailed explanations of these accidents may be found in Sehgal [14]. A short summary of each event is provided below.

The Windscale nuclear accident occurred on October 10, 1957. It took place in Unit 1 of the two unit Windscale facility reactors in Cumberland (now Sellafield), England. The Windscale reactors were graphite-moderated, air-cooled reactors. When graphite is bombarded with neutrons, dislocations are incurred in the crystalline structure. These dislocations store energy, called Wigner energy, and build up over time. If a reactor is run at temperatures above 250 °C, the graphite continually anneals and there is no Wigner energy buildup. With lower temperature reactors, such as Windscale, the Wigner energy is accumulated until a release is triggered. For the Windscale accident, a Wigner energy release started a fire in the graphite. The fire burned for 2 days releasing fission products. Most of the fission products were contained via the high efficiency filters, but noble gasses including xenon were released to the environment.

The Three Mile Island nuclear accident occurred on March 28, 1979. The reactor was an 880 MWe B&W pressurized water reactor located in Middleton, PA, USA. The event started with a blockage in a condensate polisher used for cleaning the secondary loop water. This blockage caused the feedwater pumps, condensate booster pumps, and condensate pumps to turn off. This caused the trip of the turbine along with a trip of the reactor. With reduced flows and a lack of energy removal from the turbine, primary system pressure increased causing the opening of a pilot-operated relief valve (PORV). The reactor system operated as designed through these steps. The problem started with the PORV valve did not close as it should have. This caused a small leak in the primary reactor system. This leak led to a reduction in coolant inventory. Safety systems automatically initiated to inject water at high pressures within less than 1 min from the time of the PORV closure failure. Reactor operators did not immediately understand the reason for the high-pressure water injection, so they throttled it back. Eventually, the reactor vessel water level dropped to below the core level. Lack of heat removal caused temperatures to rise, oxidation of the zircaloy cladding, and eventual fuel damage. Fission products were released into the reactor vessel and reactor containment. These fission products were largely contained, but a containment venting to reduce pressure did allow noble gasses and some volatile fission products to escape.

The Chernobyl nuclear accident occurred on April 26, 1986 (see Chap. 2). The reactor was a RBMK-1000 pressure tube graphite-moderated reactor in Pripyat, Ukraine. Preceding the accident, the reactor was being set up to run a safety test that measured the power output from the turbine as it coasted down for shutdown. The reactor conditions at the time of the test were unstable due to ^{135}Xe buildup ("xenon poisoning") in the core and a positive void coefficient of reactivity due to the reactor design. A power transient was followed by two explosions. The first explosion was likely a steam explosion and the second explosion was likely a hydrogen detonation. The reactor fuel integrity was compromised, and fission products as well as fuel particles had a direct path to the environment. With the high temperatures and ample supply of oxygen, the graphite moderator ignited into a fire. The fire continued for 9 days as fission products were lofted into the atmosphere.

The Fukushima Daiichi nuclear accident occurred on March 11, 2011 (see Chap. 2). The accident included Units 1, 2, and 3. Unit 1 was a BWR-3 design with a rated power of 460 MWe. Units 2 and 3 were BWR-4 designs with a rated power of 784 MWe. Units 4, 5, and 6 were shut down for maintenance and subsequently their cores were not damaged as part of the event. The Fukushima Daiichi nuclear accident was initiated by a tsunami triggered by the 9.0-M_W Tōhoku earthquake. The tsunami wave was 13–15 m in height and surpassed the 5.7-m seawall. While the reactors were shut down due to the earthquake, they still required decay heat removal via forced convection. The water from the tsunami eliminated the off-site power, on-site emergency diesel generators, and on-site back-up electrical batteries. As a result of the station blackout, heat was not removed from the reactors and temperatures rose. Eventually, the water levels within the reactor systems dropped below the core and fuel damage ensued. Hydrogen, produced from zirconium oxidation, produced detonations that caused structural damage. With fuel damage along with containment building damage from the detonations, fission products had a pathway into the environment.

Table 16.4 shows the ^{133}Xe release magnitude and release fractions from the four accidents of interest. All of these accidents included reactor core damage. Release magnitudes range from Windscale with 1.20×10^{16} Bq released to Fukushima Daiichi with 1.14×10^{19} Bq released. Release fractions from the reactor accidents range from 63% to 100%. Variations in scale are a function of the core size, power history, severity of core damage, reactor containment, and the number of reactors involved in the accident.

Table 16.4 ^{133}Xe release magnitude and release fraction for Windscale, Three Mile Island, Chernobyl, and Fukushima Daiichi accidents

Nuclear accident	^{133}Xe released (Bq)	^{133}Xe release fraction (%)
Windscale (1957)	1.20×10^{16} [a]	100[b]
Three Mile Island (1979)	5.85×10^{16} [c]	93[c]
Chernobyl (1986)	6.51×10^{18} [d]	100[d]
Fukushima Daiichi (2011)	1.14×10^{19} [e]	92[e]

[a]Clarke [15]
[b]Loutit et al. [16]
[c]Battit and Peterson [17]
[d]Métivier [18]
[e]Eslinger et al. [19]

The magnitude of these radioxenon releases did not impose major health concerns to the public, but they were well within a detectable range at locations across the globe. Release of other fission products including ^{90}Sr, ^{131}I, and ^{137}Cs may be of magnitude for potential health concerns, but health effects are beyond the scope of this chapter. Other radioxenon isotopes including ^{131m}Xe, ^{133m}Xe, and ^{135}Xe were also released from these accidents as they are all produced with relatively high yield from fission. However, release values of ^{131m}Xe, ^{133m}Xe, and ^{135}Xe were not available for all of these events.

16.4 Radioxenon to Aid in Nuclear Accident Response

Atmospheric radioxenon activity concentrations are routinely monitored around the world with a focus on nuclear non-proliferation. The International Monitoring System (IMS) was set up to monitor the world for atmospheric radionuclide concentrations as part of the Comprehensive Nuclear Test-Ban Treaty (CTBT) [20]. The IMS has an international network of radionuclide monitoring stations that include the ability to measure radioxenon atmospheric activity concentrations at levels below 1 mBq m^{-3}. The purpose of the radioxenon measurement capabilities within this network is to detect clandestine nuclear weapon tests to detect non-compliance with the CTBT by signatory nations. However, the capabilities of this detection network along with technology advancements made possible through the IMS develop facilitate capabilities to aid with the nuclear reactor emergency response.

16.4.1 Sensitive Detection of Nuclear Fuel Damage

The IMS radioxenon monitoring capability was not available for the Windscale, Three Mile Island, and Chernobyl nuclear accidents. However, the network was available to provide important information regarding the Fukushima Daiichi nuclear accident. First radioxenon measurements were made in the USA starting 4 days after the Tōhoku earthquake and the detections continued for weeks after the accident [21]. Aerial and ground-based measurements for radioxenon following the Fukushima Daiichi nuclear were also made in Canada [22, 23]. Detections followed the Fukushima radioxenon plume as it traveled across the Northern Hemisphere [24]. ^{133}Xe activity concentrations were measured as high as 17 Bq m^{-3} in Ashland, KS, USA. While these activity concentrations are well below any health concerns, they are a factor of 17,000 above the detection limit system requirement for radioxenon monitoring systems within the CTBT. These results may be extrapolated to demonstrate that the current radioxenon IMS has the capability to measure a 0.01% radioxenon inventory release from a reactor approximately 10,000 km away. As such, radioxenon monitoring is very sensitive and can measure minute fuel breaches in a nuclear reactor's fuel.

16.4.2 Quantification of Release Magnitude and Fuel Damage

In addition to detecting a nuclear fuel radioxenon release, radioxenon atmospheric concentrations when combined with atmospheric transport calculations may be utilized to quantify the magnitude of reactor core damage. As a noble gas, radioxenon transporting through the atmosphere is insensitive to chemical fractionation and subsequent removal. This makes radioxenon a more reliable tool for fuel release quantification in comparison to volatile radionuclides including ^{137}Cs and ^{131}I that may also be released in a nuclear accident. Eslinger et al. [19] utilized HYSPLIT [25] via an iterative method to match the measured radioxenon activity concentrations measured at five sampling locations across the USA and Canada. These results were able to estimate the radioxenon release of 1.14×10^{19} Bq of ^{133}Xe released from the accident consisting of 92% of the core inventory of Units 1, 2, and 3.

16.4.3 Accident Progression and Recriticality

Continual monitoring of the radioxenon atmospheric activity concentrations provides insight into the accident progression. In the aftermath of the Fukushima Daiichi nuclear accident, there were concerns regarding the recriticality of the core [26]. Two independent radioxenon methods may be utilized to assess if the reactor core remains in a shutdown condition.

The first method utilizes the radioxenon fission products. Figure 16.2 shows the ^{133}Xe/^{131m}Xe activity ratio as measured in Ashland, KS, USA, following the Fukushima Daiichi nuclear accident [27]. These data are compared to an ORIGEN ARP [28] model that calculates the activity ratio assuming reactor shutdown at the time of the Tōhoku earthquake. The ^{133}Xe/^{131m}Xe activity ratio serves as a chronometer that starts at the time of reactor shutdown. The data show consistency between model and the measurements in Ashland, KS. This builds confidence in the theory

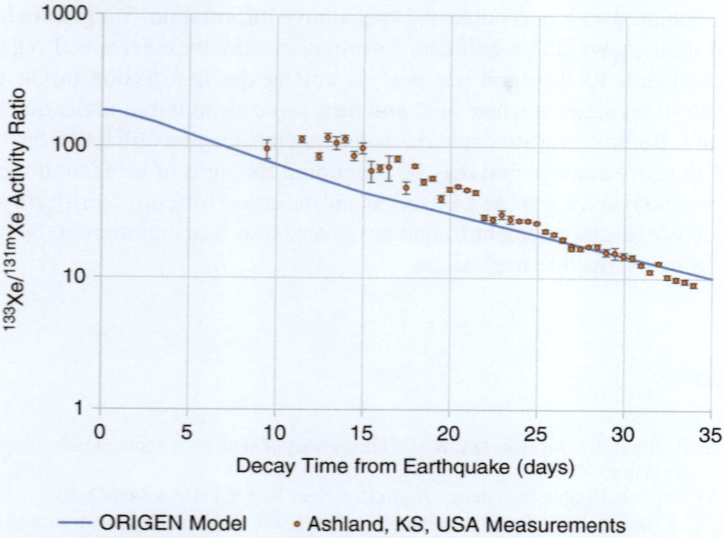

Fig. 16.2 ^{133}Xe/^{131m}Xe activity ratio measured at Ashland, KS, USA, International Monitoring System station following the Fukushima Daiichi nuclear accident

that the reactor shutdown at the time of the earthquake and a significant number of fission events did not occur after that time.

The use of fission products to determine recriticality is useful, but it may be limited to the significant levels of fission products present in the core at shutdown. Detection of new fission products produced from a very low power criticality even would be limited as the signal from the new criticality would have to exceed the background that is already present. A second method utilizes activation of xenon in air as another indicator. Xenon in air is activated in the presence of neutrons as would be present from a criticality event in a severe nuclear accident. Examination of the stable xenon isotopes shows that ^{124}Xe has a large radiative neutron capture cross-section of 146.7 b or the Maxwellian average at 0.0253 eV as shown in Table 16.1. The resonance integral radiative capture cross-section for ^{124}Xe is 2968 b. These large capture cross-sections indicate that ^{125}Xe is a strong indicator for the neutron activation of xenon in air [29]. Presence of ^{125}Xe in the atmosphere following a nuclear accident indicates that air had been activated by neutrons. This could be the result of a recriticality after the structural integrity of a nuclear reactor system had been compromised.

16.5 Conclusion

During the initial response to a nuclear reactor accident, all reliable avenues of information should be pursued. History has shown that there is significant conjecture during this time period regarding the severity of the accident and that the media is often populated with unreliable and speculative information. Study of radioxenon emission data shows that significant information may be determined regarding a nuclear accident. Radioxenon gas will be among the first fission products to be released from damaged nuclear fuel, and they serve as an initial indication of cladding failure. Radioxenon atmospheric measurements may be utilized to quantify the extent of the core damage and may be monitored for signs of accident progression. The radioxenons may also be utilized as an indicator for core recriticality. While severe nuclear reactor accident frequency is very low, it is important to be prepared for any future events that may occur.

References

1. Hwang SC, Lein RD, Morgan DA. Kirk-Othmer encyclopedia of chemical technology. 5th ed. New York: Wiley; 2005.
2. Moe M, Vogel P. Double beta decay. Annu Rev Nucl Part Sci. 1994;44:247–83.
3. Singh S, Chandra R, Rath PK, Raina PK, Hirsch JG. Nuclear deformation and the two-neutrino double- β decay in ^{124,126}Xe, ^{128,130}Te, ^{130,132}Ba and ^{150}Nd isotopes. Eur Phys J A. 2007;33(4):375–88.
4. Saey PRJ. The influence of radiopharmaceutical isotope production on radioxenon background. J Environ Radioact. 2009;100:396–406.
5. Hebel S. Genesis and equilibrium of natural lithospheric radioxenon and its influence on subsurface noble gas samples for CTBT on-site inspections. Pure Appl Geophys. 2010;167:463–70.
6. Wilson WH, Johnson CM, Lowrey JD, Biegalski SR, Haas DA. Cosmic-ray induced production of radioactive noble gases in the atmosphere, ground, and seawater. J Radioanal Nucl Chem. 2015;305(1):183–92.
7. Johnson C, Biegalski S, Haas D, Lowrey J, Bowyer T, Hayes J, Suarez R, Ripplinger M. Detection in subsurface air of radioxenon released from medical isotope production. J Environ Radioact. 2017;167:160–5.
8. Hobbins RR, Petti DA, Osetek DJ, Hagerman J. Review of experimental results on light water reactor core melt progression. Nucl Technol. 1991;95:287–307.
9. Konings RJM, Wiss T, Gueneau C. In: Morss LS, Fuger J, Edelstein NM, editors. The chemistry of actinides and Transactinide elements, vol. 6. 4th ed. New York: Springer; 2010. p. 3665–812.
10. Olander DR. Fundamental aspects of nuclear reactor fuel elements. Springfield: National Technology Information Services; 1976.
11. Iglesias FC, Lewis BJ, Reid PJ, Elder P. Fission product release mechanisms during reactor accident conditions. J Nucl Mater. 1999;270(1):21–38.
12. Andre B, Ducros G, Leveque JP, Osborne MF, Lorenz RA, Maro D. Fission product releases at severe light water reactor accident conditions: ORNL/CEA measurements versus calculation. Nucl Technol. 1996;114:23–50.
13. Konings RJM, Wiss T, Benes O. Predicting material release during a nuclear reactor accident. Nat Mater. 2015;14:247–52.
14. Sehgal BR. Nuclear safety in light water reactors, sever accident phenomenology. Waltham: Elsevier; 2012.
15. Clarke RH. An analysis of the 1957 Windscale accident using the WEERIE code. Ann Nucl Sci Eng. 1974;1(2):73–82.
16. Loutit JF, Marlay WG, Russel RS. The hazard to man of nuclear and allied radiations. 2nd report to the Medical Research Council, appendix H. London: HMSO; 1960.

17. Battist L, Peterson HT Jr. Radiological consequences of the three mile island accident, office of standards development. Washington: U.S. Nuclear Regulatory Commission; 1980.
18. Métivier H. Chernobyl: assessment of radiological and health impacts, 2002 update of Chernobyl: ten years on. Paris: Nuclear Energy Agency Organisation for Economic Co-Operation and Development; 2002.
19. Eslinger PW, Bowyer TW, Cooper MW, Haas DA, Hayes JC, Miley HS, Rishel JP, Woods VT, Biegalski SR, Hoffman I, Korpach E, Yi J, Ungar K, White B. Source term estimation of radioxenon released from the Fukushima Dai-ichi nuclear reactors using measured air concentrations and atmospheric transport modelling. J Environ Radioact. 2014;127:127–32.
20. Bowyer TW, Schlosser C, Abel KH, Auer M, Hayes JC, Heimbigner TR, McIntyre JI, Panisko ME, Reeder PL, Satorius H, Schulze J, Weiss W. Detection and analysis of xenon isotopes for the comprehensive nuclear-test-ban treaty international monitoring systems. J Environ Radioact. 2002;59(2):139–51.
21. Bowyer TW, Biegalski SR, Cooper M, Eslinger PW, Haas D, Hayes JC, Miley HS, Strom DJ, Woods V. Elevated radioxenon detected remotely following the Fukushima nuclear accident. J Environ Radioact. 2011;102(7):681–7.
22. Sinclair LE, Seywerd HCJ, Fortin R, Carson JM, Saull PRB, Coyle MJ, Van Brabant RA, Buckle JL, Desjardins SM, Hall RM. Aerial measurement of radioxenon concentration off the west coast of Vancouver Island following the Fukushima reactor accident. J Environ Radioact. 2011;102(11):1018–23.
23. Zhang W, Ungar K. Atmospheric particulate and xenon radioisotopes monitoring over Canada and health impact estimation before and after the Fukushima nuclear accident. Radiat Prot Dosim. 2017;79(2):136–41.
24. Stohl A, Seibert P, Wotawa G. The total release of xenon-133 from the Fukushima Dai-ichi nuclear power plant accident. J Environ Radioact. 2012;112:155–9.
25. Draxler RR, Hess GD. An overview of the HYSPLIT_4 modelling system of trajectories, dispersion, and deposition. Aust Meteorol Mag. 1998;47:295–308.
26. Darnowski P, Potapczyk K, Świrski K. Investigation of the recriticality potential during reflooding phase of Fukushima Daiichi Unit-3 accident. Ann Nucl Energy. 2017;99:495–509.
27. Biegalski SR, Bowyer TW, Eslinger P, Friese JA, Greenwood LR, Haas DA, Hayes JC, Hoffman I, Keillor M, Miley HS, Moring M. Analysis of data from sensitive U.S. monitoring stations for the Fukushima Daiichi nuclear reactor accident. J Environ Radioact. 2012;114:15–21.
28. Oak Ridge National Laboratory (ORNL). ORIGEN: origen-ARP2. Oak Ridge: ORNL; 2002.
29. Klingberg FJ, Biegalski S. Radioxenon signatures by natural activation. J Radioanal Nucl Chem. 2013;296(1):117–23.

Chapter 17
Response to Nuclear Terrorism in Germany

Britta Lange

Abstract This chapter describes the responsive approach to nuclear terrorism in Germany. Germany is committed to responding to all kinds of criminal acts involving nuclear and other radioactive material out of regulatory control through international treaties and national law.

After a short introduction and a definition of nuclear terrorism, particular attention will be paid to Germany's international commitment and national strategies. Due to the scope and complexity of the task at hand and the large number of institutions involved, this chapter shall focus on the practical approach to the response to nuclear terrorism in Germany, especially in regard to radiation protection, and will be limited to the responsibility of the federal authorities. An exhaustive representation of German capabilities was not attempted, as this would go beyond the scope of this publication.

Keywords Nuclear terrorism · Nuclear security event · Misuse of radioactive material · Radiation detection · Central Federal Support Group for the Response to Serious Nuclear Threats (ZUB) · Mobile Expert Support Team (MEST) · Radiological crime scene work · Nuclear forensics analysis

17.1 Introduction

Terrorism remains a threat to international stability and security. A special type of terrorism is nuclear terrorism, which is defined extensively in Article 2 of the United Nations International Convention for the Suppression of Acts of Nuclear Terrorism [1]. According to this article, nuclear terrorism is an offense committed if a person unlawfully and intentionally possesses or uses radioactive material or a nuclear device or damages a nuclear facility with the intent to cause death or serious bodily injury or with the intent to cause substantial damage to property or to the

B. Lange (✉)
RN 7 Response to Nuclear Security Events, Bundesamt für Strahlenschutz – German Federal Office for Radiation Protection, Berlin, Germany
e-mail: blange@bfs.de

G. Steinhauser et al. (eds.), *Nuclear Emergencies*,
Current Topics in Environmental Health and Preventive Medicine,
https://doi.org/10.1007/978-981-13-8327-4_17

environment or with the intent to compel a person, an organization, or a State to do or refrain from doing an act. This means nuclear terrorism is not limited to the use of nuclear devices. Any use of radioactive material with the intent to cause serious damage can be seen as nuclear terrorism. The threat of nuclear terrorism has been recognized as a matter of concern for all States, and the risk that nuclear material or other radioactive material could be used in a criminal act remains high and is regarded as a serious threat to international peace and security. Therefore, it is vital that each State establishes an appropriate and effective nuclear security regime to combat nuclear terrorism.

There are large quantities of radioactive material in existence, which have applications in various fields, such as health, the environment, agriculture, and industry. The hazards associated with this material vary according to composition and activity. Additionally, the use of explosives in combination with this material can drastically enhance the impact of a criminal or terrorist act. If a criminal or terrorist group managed to detonate a radiological dispersal device (RDD) or often called "dirty" bomb in an urban area, the result could be mass panic, widespread radioactive contamination, and major economic and social disruption [2, 3].

17.2 International Commitment

The threat of nuclear and radiological terrorism remains on the international security agenda. Nevertheless, to reduce this risk, the international community has made great progress in securing nuclear and other radioactive material that could otherwise be used in a terrorist act. This progress is contingent on the efforts of all States, including Germany, to adopt strong nuclear security systems and measures [4].

Germany has already committed itself to taking all necessary measures to combat nuclear terrorism. It signed international conventions, such as the Convention on the Physical Protection of Nuclear Material [5] and the International Convention for the Suppression of Acts of Nuclear Terrorism [1]. Germany signed and fully supports the Code of Conduct on the Safety and Security of Radioactive Sources [6]. Germany is a Member of the G7 Global Partnership against the Spread of Weapons and Materials of Mass Destruction and signed the United Nations Security Council resolutions on the non-proliferation of weapons of mass destruction. Germany is a partner nation of the Global Initiative to Combat Nuclear Terrorism, supported the Nuclear Security Summit Communiqués and contributed to the IAEAs Nuclear Security series. This international commitment requires that Germany has a national strategy and appropriate measures in place to handle nuclear security events.

In addition to the international involvement mentioned above, Germany signed agreements with other countries on cooperation and assistance in addressing a terrorist threat in the radiological and nuclear fields in order to establish mutual cooperation between the participants in the scientific, technical, and operational fields.

17.3 National Strategies

In order to fulfill the international obligations set out in the last paragraph and as part of the national action plan Germany has in place, among other things, strategies for the safety and security of its nuclear facilities, including cyber security, strategies for gaining control over orphan sources, for the intervention in the event of a malicious act involving a radioactive source, as well as arrangements for the appropriate training of staff in this field. Many of these measures are set out in the Atomic Energy Act and in the Radiation Protection Act [7], which came into force in 2017. The Radiation Protection Act regulates, among other things, the responsibilities of the Federal and State ("Länder") authorities for nuclear emergencies.

A regular evaluation of the nuclear threat level in Germany, which is carried out by the Federal Criminal Police Office (BKA) and other federal authorities, and a regular exchange of information within Germany, is the basis for an adequate preparation. There are reporting mechanisms besides police reporting, for instance, reporting incidents with nuclear or radioactive materials to the competent federal authorities. In regard to the security of sealed sources, Germany has a very well-maintained centralized register of Highly Active Sealed Sources (HASS), which easily enables accounting of such sources.

For an appropriate and timely response to a nuclear security event, a range of specialists are required and need to be included in the planning and prevention stage: police, radiation protection, and environmental protection officers from all levels of government (from forces deployed at the scene to those working in the ministries) and from other institutions such as universities and laboratories. These specialists can come from the Länder, the federal government, and from the EU or beyond. With sixteen Länder together with the federal government and with many different active partners, Germany is faced with a considerable cooperation and communication challenge for the response to nuclear security events. However, the advantage of such a large active network is its natural resilience.

Each state ("Land") is directly responsible for the first response to all emergency situations and for the police response on its own territory, including, generally, the response to nuclear security events, as set out in German Basic Law. Since every Land has different ways of dealing with nuclear hazards, different regulations, and specialists, it is crucial for federal forces to retain a high degree of flexibility in order to work together effectively. Coordination is usually carried out by the federal government. In the event of a serious emergency with radioactive material, the German Land concerned can receive support from the federal government in the form of additional forces to deal with the event. The federal government has a Mutual Crisis Committee in place formed by the Federal Ministry of the Interior, Building and Community (BMI) and the Federal Ministry for the Environment, Nature Conservation, and Nuclear Safety (BMU) for such cases.

In the event of an act of terrorism, the responsibility can completely or partially be handed over to the federal authorities. Support for the competent authority, state or federal, can be in the form of administrative assistance by a particular federal institution (e.g., by the Federal Office for Radiation Protection (BfS) or the BKA) or, in the event of the criminal use of radioactive material, by the joined forces of the Central Federal Support Group for the Response to Serious Nuclear Threats (abbreviated to "ZUB" from the German term), which is an ad hoc response support group that brings together the BKA, the Federal Police (BPOL), and the BfS.

17.4 The German Federal Office for Radiation Protection (BfS)

The BfS is a scientific and technical federal office with a focus on the measurement and control of radiation and the safety and security of the use of all kinds of radiation (ionizing and non-ionizing). Nuclear emergency preparedness and response at the BfS and supporting other authorities during the response to nuclear security events is written into law. The task involves preparing for and responding to situations in which radioactive or nuclear material is out of regulatory control. That can be the loss or discovery of radioactive materials or sources, including risk assessment, or cases where the material could be or is being used maliciously. The BfS can be called upon to support other German authorities (on the federal level or from the Länder) if the authority responsible is not able or does not have the capacity to respond itself. The tasks performed by the BfS will vary depending on the type of deployment and the resources and capabilities of the requesting German authority. The BfS can support with advice in all fields of radiation protection and measurement, with trained personnel who can operate in the field and with specialized radiological measurement equipment, including remotely controlled equipment (Fig. 17.1).

The BfS has various kinds of specialized Mobile Expert Support Teams (MESTs): for the characterization of radioactive and nuclear material (RN-material), for covert searches for RN-material, for open searches for RN-material on the ground, and for

Fig. 17.1 Remote-controlled caterpillar vehicle with high-purity germanium detector for the identification of radioactive material ("manipulator robot," exercise scenario)

open searches for RN-material from the air. Additionally, BfS staff can carry out dispersion calculations after a release of radioactive material into the environment, e.g., after an explosion, including the related radiation exposure to the deployed forces and population. Usually, BfS works as one subunit of the ZUB during the response to nuclear security events. The main task of the BfS in the ZUB is the radiological threat assessment including detection and identification of radioactive material. The BfS advises the BKA and the BPOL, as well as other police authorities, on the radiation protection measures and the nuclear forensics possibilities in real time. The BfS aims to apply measures to allow normal police work to continue within a reasonable timeframe, despite the presence of nuclear or other radioactive material out of regulatory control.

17.5 The Central Federal Support Group for the Response to Serious Nuclear Threats (ZUB)

The ZUB is formed as part of the response to a nuclear security event, in particular to criminal acts involving nuclear or other radioactive materials out of regulatory control. It is a cooperation between the BKA, the BPOL, and the BfS. If support from the ZUB has been requested by a German Land, then the ZUB will join the police investigation as a subsection, with the police commander from the Land remaining in charge of the entire operation.

The ZUB has specialized subunits and is trained to work in the presence of RN-material. The ZUB can provide teams for surveillance and covert detection, meaning the localization of radioactive sources through the use of covert measuring equipment, for instance, backpacks and vehicle-based detection. It has Special Weapons and Tactics (SWAT) teams formed by the Border Protection Group 9 (GSG9) of the BPOL, who are trained and equipped to work in contaminated areas (Fig. 17.2).

The subunits of the ZUB can perform wide area searches for radioactive sources including foot- and vehicle-based measurements and airborne gamma spectroscopy measurements. There are bomb disposal experts, as well as experts for crime scene work and for decontamination and medical care. Additionally, the ZUB has its own forces for air transport and logistics.

In the event that a suspicious object is found, the bomb disposal experts of the ZUB make sure that it is safe to approach, collect basic radiological data, perform remote analyses, and assess the situation. Additionally, neutron- and gamma-measurements are performed including the detection of fissionable materials and evaluation of criticality risks. If necessary, a containment is set up in order to minimize the possible release of radioactive material and render-safe procedures are performed.

ZUB teams are also trained for radiological crime scene work (Fig. 17.3). At its most basic level, a radiological crime scene is a location where either a criminal act

Fig. 17.2 SWAT team members undergo contamination control measurements after deployment (during an exercise of the ZUB)

Fig. 17.3 A mixed team of police and radiation protection specialists work together in a radiologically contaminated crime scene (during an exercise of the ZUB)

involving nuclear or other radioactive material has taken place or is suspected to have taken place, or it is a location where traces or evidence related to such an act have been found [8]. It could be an area where someone tried to manipulate sealed sources, construct a dirty bomb, or the scene after a terrorist attack that intended to deliberately disperse nuclear or other radioactive material in a populated area. ZUB staff ensure that all actions at the radiological crime scene are carried out in a way that maintains the integrity of the criminal investigation and that all relevant criminal investigative procedures are applied through effective radiological crime scene management. Managing the radiological crime scene is also an integral part of law enforcement investigations and is necessary to support any future legal proceedings in relation to the nuclear security event.

Evidence contaminated with nuclear or radioactive material can undergo limited initial forensic analysis at the scene. Under usual circumstances, forensic analysis would take place in a police laboratory; however, further forensic analysis of evidence contaminated with nuclear or radioactive material is not possible in a police forensic laboratory in Germany, due to the fact that police laboratories are neither licensed nor capable of handling radioactive samples. In such an event, a close cooperation between police and radiation protection experts is needed to investigate such samples by means of nuclear forensics.

17.6 Nuclear Forensic Investigations

Nuclear forensics includes not just the forensic analysis of nuclear material, or evidence contaminated with nuclear material, but also the forensic analysis of radioactive materials and evidence contaminated with radioactive material. The analysis of nuclear or other radioactive material seeks to identify what the materials are, how, when, and where the materials were made, and what their intended uses were [9]. Nuclear forensics on radioactive materials can be carried out by the BfS, where there is a specialized laboratory for receiving and characterizing "unknown" radioactive samples, i.e., non-nuclear samples for which the nuclide identification and the activity values have not yet been determined exactly. The BfS also has a laboratory for trace analysis and can handle and analyze contaminated food samples. In addition, measurements in the whole-body counter, biological dosimetry, and the analysis of urine samples are possible if incorporation of radionuclides is suspected. The BfS also maintains the centralized register of HASS mentioned above and can trace source numbers if requested.

In the event that nuclear material or evidence contaminated with nuclear material is removed from the scene for further forensic investigation, the items in question are transported, taking into account chain of custody and all-hazards considerations, to the Joint Research Centre (JRC) Karlsruhe of the European Commission, based on a framework agreement with the German federal government. The JRC Karlsruhe has specialized facilities for handling nuclear material and contaminated evidence. They specifically developed a glove box in cooperation with the BKA for the purpose of obtaining fingerprint analysis and DNA samples from evidence contaminated with nuclear or other radioactive material. During the analysis, the police experts can direct the working steps carried out by the JRC personnel. In addition, the JRC Karlsruhe can perform nuclear forensics analysis on nuclear materials (bulk or trace) and compare the results of the measurements with their extensive database. The work of the JRC Karlsruhe is carried out within the framework of a nuclear forensics investigation plan, which is drawn up in close cooperation with the police. The aim is to produce a scientific report that can be used in the police investigation and also in court. In conclusion, it can be stated that Germany is very active in the field of nuclear forensics and that cooperation between police experts and radiation protection specialists exists at the Länder-level as well as the federal level and internationally. The cooperation involves many different police experts and radiation protection specialists and is extensively tested through exercises and deployments.

17.7 Training and Exercises

Since the ZUB has a legal obligation to support the competent authorities during the response to radiological or nuclear threats, the ZUB has to ensure its operational readiness. The three institutions of ZUB, the BfS, the BKA, and the BPOL, are very different in nature, both culturally and with respect to their expertise. In order to ensure that the forces work well together, despite the possibly difficult and varying conditions of a deployment, regularly planned exercises and training are essential for improving cooperation, capabilities, and resilience. The ZUB has a training schedule that includes both training and exercises internally within the ZUB (between BPOL, BKA, and BfS) on two different levels. The first level is training and exercises within one of the partner institutions, organized by that institution for its employees alone. The second level of training and exercises internally within the ZUB occurs between the subunits of the different institutions, for example, crime scene work exercises between forensic experts and radiation protection specialists, training police in how to use specialized radiation protection equipment and training BfS staff on police procedures and equipment.

The first level of training and exercises within each partner institution maintains the skills base, forges stronger communication links, and strengthens the feeling of commitment of the institution's members. The second level of training, between subunits of different ZUB partner institutions, strengthens the communication link between the officers in the subunits, keeps skills updated, and allows for the boundaries of the expertise of the other institution to be assessed. This is particularly useful for a deployment, as deployment leaders need to have a realistic idea of the capabilities of each institution within the ZUB.

Another type of exercise is that between the ZUB and one of the German Länder, known as a "ZUB full scale exercise." It is the largest and most costly of all the exercises and training undertaken by the ZUB, and it is arguably the most important. The lessons learned from such an exercise have a profound effect on the future course of the ZUB and many changes are made following a review of an exercise.

Exercises and training are one of the main methods used for improving best practice in the field of the response to nuclear terrorism in Germany. Exercises undertaken in a controlled manner can highlight problems, which can then be tackled in a systematic fashion to develop new operating procedures and guidelines. These guidelines and procedures are then implemented and tested in a similar fashion, yielding what is hopefully a continuous improvement in the ZUB operation.

17.8 Radiation Detection at Major Public Events

It is widely acknowledged that there is a substantial threat of a terrorist attack on major public events, such as high-profile political or economic summit meetings or major sporting contests. High-profile major public events occur regularly, capturing great public interest and receiving intense media coverage. Criminal or terrorist acts involving nuclear or other radioactive material at any major public event could result in severe consequences, depending upon the nature and quantity of the specific material involved, the mode of dispersal (violent or non-violent), the location and the population impacted. Implementing nuclear security systems and measures is, therefore, of paramount importance [3].

In Germany, the BKA is usually responsible for the security measures at such events. The strategy and scope of the security measures are based on the threat and risk assessment which is carried out by the competent authorities and will be adapted to the specific situation. If the police authorities deem it necessary, a special CBRN (chemical, biological, radiological, and nuclear defense) team consisting of experts from federal institutions can be called upon to support the security arrangements at major public events.

For the RN part, the BfS can support a requesting German security authority during a major public event [10]. The tasks of the radiation protection personnel will be communicated before the deployment and included as part of the deployment plans drawn up by the security authority. A concept of operations that combines radiation detection with metal detection is usually implemented. Procedures are put in place for the event that the presence of RN-material is detected and/or confirmed. The method of deployment of the radiation detection capability varies depending on the needs of the requesting security authority and can encompass elements such as search teams for the detection of nuclear and other radioactive material inside and outside, setting-up and running mobile radiation portal monitors for gamma and neutron radiation, providing advice when there is a threat to use radioactive material, and the analysis of samples. If samples are taken from the search area (e.g., swipe samples), then these samples are analyzed off-site (outside the search area). This can be done either by mobile equipment or in a standard lab. Labs are used if the analysis of samples is a planned part of the BfS assistance, or if it becomes necessary during the deployment to confirm results in the lab (e.g., alpha spectroscopy, liquid scintillation counting). A regular part of the security measures at major events is the analysis of food samples taken by BKA food chemists. Such samples are scanned for chemical, biological, and radiological contaminants. The BfS has developed procedures for the very rapid analysis of alpha, beta, gamma, and neutron radiation in food.

17.9 Incidents

Since the reunification of Germany, several high-profile criminal investigations involving nuclear or other radioactive material have been reported in the international press. Examples include the prominent "plutonium affair" in 1994, during which approximately 360 g of plutonium were smuggled into Munich airport, the "WAK plutonium theft" in 2000, during which contaminated material was stolen from a reprocessing facility in Karlsruhe and the "Po-210 case" in 2006, during which traces of ^{210}Po were discovered at various locations in and around Hamburg. The latter case was the first deployment of the ZUB.

In late 2006, Alexander Litvinenko died as a result of a poisoning with polonium-210 (Po-210 or ^{210}Po), which allegedly occurred at a meeting in London. Media reports at the time linked Dimitri Kovtun to this meeting and to the German city of Hamburg. An investigation was started by Hamburg Police into Kovtun's movements during a visit to Hamburg in the week directly before the alleged poisoning. As the presence of ^{210}Po at the sites visited by Kovtun in Hamburg was uncertain, Hamburg Police called on the ZUB for support. The BfS was responsible for the measurement of ^{210}Po at the sites visited by Kovtun [11], the radiological evaluation of the measurements, and the radiation protection recommendations. Following a measurement for airborne contamination at the sites involved, both

Fig. 17.4 "Po-210 case" in Hamburg 2006: BfS measurement experts at work at one of the sites investigated

field and laboratory techniques were used to monitor the ^{210}Po contamination (Fig. 17.4).

The evaluation of the measurements enabled the BfS to offer effective radiation protection advice and to assist the police investigation. Fortunately, the traces of ^{210}Po found by the BfS were of little radiological relevance and the radiation protection measures taken by BfS reflected this fact.

In summary, it is fair to say that Germany has prepared a comprehensive set of measures for the response to nuclear terrorism and is prepared to combat this type of terrorism. However, especially in these times of increasing terror—more than 45 such attacks resulting in fatalities in western Europe in 2016 and 2017—and the poisoning of Sergei and Yulia Skripal in March 2018 in the UK (with its apparent analogies to the Litvinenko poisoning), it is always necessary to improve such measures in the entire field of CBRN and to train and exercise them in order to be well prepared.

References

1. United Nations. International convention for the suppression of acts of nuclear terrorism. 2005. https://treaties.un.org/doc/db/terrorism/english-18-15.pdf. Accessed June 2018.
2. BfS. Misuse of radioactive material in connection with an unconventional explosive device ("dirty bomb"). 2007. http://www.bfs.de/EN/topics/ion/accident-management/bfs/hazard-avoidance/abuse-explosives.html. Accessed June 2019.
3. IAEA. Nuclear security systems and measures for major public events. Implementing guide, Nuclear security series no. 18. Vienna: IAEA; 2012.
4. IAEA. Objective and essential elements of a state's nuclear security regime. In: Nuclear security fundamentals, Nuclear security series no. 20. Vienna: IAEA; 2013.
5. IAEA. Convention on the physical protection of nuclear material. 1979. https://www.iaea.org/publications/documents/infcircs/convention-physical-protection-nuclear-material. Accessed June 2018.
6. IAEA. Code of conduct on the safety and security of radioactive sources. Vienna: IAEA; 2004.
7. Ziegler E, Mann T. Atomrecht und Strahlenschutz. Baden-Baden: Nomos Verlagsgesellschaft; 2019.
8. IAEA. Radiological crime scene management. Implementing guide, Nuclear security series no. 22–G. Vienna: IAEA; 2014.
9. IAEA. Nuclear forensics in support of investigations, Nuclear security series no. 2-G. Vienna: IAEA; 2015.
10. Kröger EA, Maier R. Radiation detection at major public events from the perspective of the German Federal Office for Radiation Protection. In: IAEA international conference on nuclear security: enhancing global efforts. Vienna: IAEA; 2013. p. 463.
11. Kröger EA. The Litvinenko polonium-210 case-German experiences. Strahlenschutzpraxis. 2009;1:16.

Chapter 18
Nuclear Emergency Preparedness in Germany: Lessons Learned from Fukushima and Chernobyl and Their Implementation

Matthias Zähringer and Florian Gering

Abstract Both the Chernobyl and Fukushima nuclear reactor disasters had a huge impact on nuclear energy policy in general and in particular on nuclear emergency planning and response. The Chernobyl accident demonstrated the need for additional planning for precautionary protective actions on food production on a large scale as well as for international harmonization of emergency preparedness and response. As a consequence, the German Integrated Measurement and Information System (IMIS) was established. The Fukushima Daiichi NPP accident demonstrated not only the possibility of severe beyond design accidents in western-type light water reactors but also highlighted the need for in-depth preparation of protection strategies, in particular for INES-7-category accidents, and the outstanding role of psychosocial issues in disaster mitigation and recovery. Following guidance of international bodies and the European directive 2013/59/EURATOM, the Federal Government of Germany reviewed and updated its entire nuclear emergency planning and response framework since 2011. Under the new German radiation protection law, a process of implementing a modern nuclear emergency management system is currently under way.

Keywords Nuclear emergency preparedness · Nuclear disaster · Protective actions · Radiation protection · Germany

---

M. Zähringer (✉) · F. Gering
Emergency Preparedness Division, Federal Office for Radiation Protection, Freiburg, Germany
e-mail: mzaehringer@bfs.de

G. Steinhauser et al. (eds.), *Nuclear Emergencies*, Current Topics in Environmental Health and Preventive Medicine, https://doi.org/10.1007/978-981-13-8327-4_18

18.1 Introduction

Emergency preparedness and response (EPR) for severe nuclear reactor accidents and other nuclear or radiological disasters is a complex issue involving numerous disciplines, including nuclear physics and engineering, chemistry, meteorology, biology, medicine, psychology, social sciences, and economy. Managing a nuclear accident or disaster has many challenges in common with managing other non-nuclear disasters such as flooding or hurricanes. However, protection strategies are often discussed and developed by radiation protection experts only. Maintaining the appropriate level of interdisciplinary exchange is a requirement yet to be fulfilled.

Another specific problem of EPR is that—luckily—nuclear disasters are extremely seldom. As a consequence, most countries plan for situations they never had to and probably never will have to manage. In order to ensure that EPR is well established, there are a number of tools to employ: Using experience and cross fertilization from non-nuclear disaster management, exercises, evaluation and quality assurance, scenario development, and, most importantly, a careful analysis of the few nuclear disasters mankind has experienced so far.

18.2 Post-Chernobyl EPR in Germany

In May 1986, there was no overarching federal plan in Germany on how to cope with a large-scale cross-boundary contamination. Existing monitoring programs and planning for protective actions were linked to domestic nuclear facilities and the accountability was with the licensing federal state. Provisions were in force to protect the population against severe adverse effects from reactor accidents up to a regional scale. However, there was no concept on how to reduce radiation doses in regions where no hard disaster protection measures were needed but doses due to ingestion and external exposure could be well detectable and in the order of a few millisievert. In particular, there were:

- No program and capabilities to monitor the full territory of Germany in sufficient geographical resolution for ambient dose rate, air activity concentration, and activity concentration in other environmental media and in the food production chain;
- No capabilities to aggregate and disseminate quality controlled measurement data nationwide in standardized formats;
- No permissible limits for activity concentrations in food and feedstuff;
- No concept on how to limit and reduce the dose to the public at relatively low levels in a balanced and justified way and no plan for appropriate protective actions to apply; and
- No communication concept to recommend in plain language reasonable protective actions and to explain why others make no sense.

It became apparent that radioactive material can be transported over large distances at concentrations that may have certain impact on the population and requires some action. Particularly, safety of food was subject to controversial debates and problems of marketing food had high economic impact.

The situation severely worsened as there was no sufficient data available to assess adequately the radiological situation and thus information given to the public by the federal government, state (Länder) governments, and NGOs was often incorrect, misleading, and even contradictory. The result was a loss of trust in government and authorities and a widespread emotional overreaction by the public.

In response to these shortcomings and misperceptions, the federal government began to plan, implement, and commission an Integrated Monitoring and Information System (IMIS), comprising networks for monitoring ambient dose rate, air concentrations and water contamination, laboratory measurement programs as well as appropriate communication and IT infrastructure [1]. This task was achieved in close cooperation with the federal states. The legal framework was defined in the new precautionary radiation protection law (StrVG) which filled the gap of additional protective actions besides the "classic" actions of disaster management, i.e., evacuation, sheltering, and iodine thyroid blocking.

With this monitoring system, Germany implemented one of the densest and most sophisticated monitoring programs in the world for environmental monitoring and nuclear emergency preparedness. The already existing dose rate monitoring network (established decades before for civil defense purposes) was technologically upgraded with more sensitive probes. With an average distance of about 20 km between probes countrywide and additional probes around nuclear installations it is now the network with the highest spatial density in Europe.

It was also acknowledged that emergency planning and response must be harmonized between neighboring countries, particularly member states of the European Community. Consequently, the EC responded with issuing maximum permissible activity concentrations in food and feeding stuff to be applied in the common European market. Common European exchange platforms for environmental radiological monitoring data (EURDEP) and early notification for nuclear accidents (ECURIE) were created.

Epidemiological studies recorded unexpected high thyroid cancer rates in children in the most affected areas in the Ukraine, Belarus, and Russia few years after the accident [2]. The German radiation protection advisory board reacted on the new risk estimations for thyroid cancer and lowered the intervention level of iodine thyroid blocking for children, adolescents, and pregnant women to 50 mSv thyroid dose. Planning zones, in which iodine thyroid blocking was planned for children, adolescents, and pregnant women, were accordingly increased from 25 km up to 100 km. Other planning radii were not changed. The particular characteristics of the Chernobyl reactor accident were too specific for RBMK type reactors and the scenario, in particular the source term, was not appropriate as a basis for a hazard assessment for nuclear power plants in Germany.

18.3 The Fukushima Shock

The Great Eastern Earthquake and Tsunami disaster in Japan and the resulting core melt in three reactors at Fukushima Daiichi nuclear power plant (NPP) in the aftermath had a great impact on the general attitude against nuclear energy in Germany in general. A few weeks later, the German federal government decided on an accelerated nuclear phase-out with the immediate shutdown of several reactors and a planned shutdown of the remaining reactors before end of 2022. With this decision, the risk of a severe domestic reactor accident is now decreasing. However, other nuclear threat scenarios, for example, nuclear accidents in foreign NPPs near the border, remain.

The Fukushima accident had no radiological impact within the German territory. Only traces of radioiodine and other fission products were detected in German monitoring systems after 23 March 2011 [3, 4]. Concentrations were of the order of few millibecquerels per cubic meter. However, the interest of the public in environmental data, particularly in data from the German dose rate monitoring, was huge and the Federal Office for Radiation Protection (BfS) had to respond accordingly: Information offers had to be hosted on scalable servers and sufficient capability to respond to a high number of requests. Data given to the public were not well understood and design and explanatory text had to be added or improved where already existing. Following an open data access policy, BfS decided to publish the data immediately after polling from the probes rather than withholding data until being fully verified. It turned out that the data quality was high enough and erroneous data did not pose a problem [5]. The analysis of public response to BfS data showed clearly that "data only" is not sufficient to satisfy the population's needs.

Fukushima was the first accident with a core melt in a western-type pressurized water reactor falling into the highest INES category 7 [6]. It happened in an industrialized democratic country comparable with other highly developed countries like Germany. As a consequence, the German Radiation Protection Commission (SSK) reviewed the EPR system in Germany based on the lessons learned in Japan and summarized experts' discussions in a publication containing 76 detailed recommendations on how to improve the EPR system in Germany [7]. The most important recommendation was to increase planning radii for precautionary and urgent protective actions around power reactors. The old and new numbers are given in Table 18.1. The new radii are based on a rigorous scientific approach for hazard assessment of nuclear accidents [8] and a clear separation between scientific findings and political decisions. The new approach started with a change of paradigm: Beyond design accidents up to INES category 7 were now explicitly taken into consideration. Any calculated or assumed low probability was no longer accepted as a cut-off criterion to not consider such extreme scenarios. It was agreed to use a few specific source terms assessed by the "Gesellschaft für Anlagen- und Reaktorsicherheit" (GRS) as new reference scenarios. They serve as basis of the hazard assessment and thus of the whole EPR system for nuclear accidents and they were used as input for extensive consequence analysis based on the decision support system RODOS. The

Table 18.1 Old and new NPP emergency planning zones

Zone	Protective action	Old radius for NPP	New radius for NPP
Central zone	Evacuation within 6 h[a]	2 km	5 km
Middle zone	Evacuation within 24 h[a]	10 km	20 km
Outer zone	Iodine thyroid blocking (all <45 years), Sheltering	25 km	100 km
Remaining national territory	Iodine thyroid blocking for children <18 years and pregnant women	100 km	Remaining national territory

[a]Sheltering and iodine thyroid blocking also apply

largest of the new reference source terms comprises a release of radioactivity equivalent to 1600 PBq iodine-131 (this compares to estimated 100–500 PBq iodine-131-equivalent from Fukushima [9] and 1760 PBq from Chernobyl [10]). Atmospheric dispersion and resulting dose to the public were calculated for three representative NPP sites in Germany and with real weather data for each site for 365 days of one reference year. According to the German legislation, specific dose criteria exist for each of the three typical precautionary and urgent protective actions: evacuation, sheltering, and iodine thyroid blocking. In the hazard assessment, those areas were identified for all weather scenarios, where those dose criteria would have been exceeded. A careful analysis of frequency and maximum distance from the source led to the abovementioned new emergency planning zones [4].

Inspired by international guidelines of the ICRP and following the EU council decision 2013/59 EURATOM, the radiation protection in general and the EPR system in particular were completely refurbished in Germany. The legislative basis was integrated and several laws and ordinances were comprised into the new radiation protection law (StrlSchG).

Most prominently, a new Federal Radiological Emergency Response Centre (RLZ) was created, which will replace several radiological emergency response centers in federal states. It is comprised of operating units in the Federal Ministry of Environment, Nature Conservation, and Nuclear Safety (BMU), the Federal Office for Radiation Protection (BfS), and the GRS. The RLZ will only be established in case of a nuclear accident and will then become an entity of BMU. Supporting organizations are connected to the center by redundant and specifically hardened communication lines. Computer systems and other equipment are highly redundant allowing operation of the center even if single location is out of operation due to major disaster consequences, which may occur in conjunction with a severe radiological emergency. This resilience against large-scale failures of infrastructure is also a consequence of the Fukushima event analysis and the observed malicious synergy of earthquake, tsunami, and nuclear disaster.

The tasks of the new Federal Radiological Emergency Response Centre are:

- Collecting data;
- Compilation of a comprehensive radiological situation report;
- Dissemination of information within German emergency response organizations;

- International information exchange;
- Coordination of protective measures among public stakeholders;
- Information of the public; and
- Coordination of radiological measurements within Germany.

18.4 An Outline of Modern EPR: The Way Forward

Detrimental consequences of the fast evacuations in Fukushima province became apparent. At least 50 elderly persons died during or directly after the evacuation [11] and over the following years a clear increase of mortality rates was reported for the evacuees [12]. On the other hand, neither radiation-related deaths nor deterministic effects caused by the increased exposure have been found up to now in Japan [9, 13]. It can be concluded that the evacuation of about 160,000 people did more harm than good to them, especially to some of the most vulnerable fractions of the population (elderly people and people suffering from diseases). This experience from Fukushima begs questions on the process of decision-making, in particular on precautionary and urgent protective actions. As frequently observed in exercises, decision-makers tend to make decisions "on the safe side." The whole process of decision-making from atmospheric dispersion prognosis to finally identifying affected municipalities has an implicit bias of being too "conservative" [14].

The protective action of evacuation is not only specific to nuclear emergencies. It is planned and carried out in case of a number of natural and man-made disasters. This also holds true for a number of other protective measures. A severe reactor accident has huge impact far beyond direct radiological consequences. Protective actions, remediation, and recovery are a complex societal process. The alignment of EPR with non-nuclear disaster preparedness is one of the big lessons learned and discussed internationally [15].

Beyond this discussion there is a deep change of the understanding of the role of radiation protection experts. They can consider only the radiological aspects of the emergency and they are only advisors for those who plan, decide on, and implement protective measures. The new German legislation consistently follows this concept, called "Verzahnungsansatz," which can be roughly translated to "interlinking approach." Emergency planning is specified in a General Emergency Plan, which is under the responsibility of the BMU, and several Special Emergency Plans focusing on specific issues such as Agriculture, Traffic, Drinking Water, and Waste Management. Each Special Plan is under the responsibility of the respective accountable Ministry. One key task of radiation protection experts in the abovementioned Federal Radiological Emergency Response Centre is to define and adjust reference levels and to advise on maximum permissible levels in environmental media, to assess and forecast where and when these levels are exceeded.

A few days after the Fukushima accident the international, the volunteer-centered "citizen science" organization "Safecast" was formed. It quickly began monitoring, collecting, and openly sharing information on environmental radiation. Its mission is "to provide citizens worldwide with the tools they need to inform themselves by gathering and sharing accurate environmental data in an open and participatory fashion" [16]. Data collected from, e.g., car borne devices were uploaded on a publicly available server and aggregated in a visual form clearly delineating affected areas. In doing so they contributed independent trustworthy information. Emergency planning and preparation should be open to such civil society contributions.

Health impact of a major nuclear accident always includes impact attributable to ionizing radiation and psychosocial health and societal impacts. In Fukushima, psychosocial and societal problems had clearly the highest impact. Germany's accelerated phase-out from nuclear energy after the Fukushima disaster was driven by a broad and deep public concern about the risk of radioactivity and nuclear energy. It is safe to say that a similar event in Germany would have even more pronounced societal and psychosocial impacts compared to Japan. The consequences of this finding will have far reaching consequences on EPR and will require a re-evaluation of basic concepts and protection strategies. A full discussion goes beyond the scope of this paper. However, we would like to outline a short but not exhaustive list of issues to be addressed in future:

– Protective measures must do more good than harm. It is unclear how to adequately apply psychosocial adverse effects in a quantitative fashion, both in considerations at the preparedness stage and in decision-making in an emergency.
– The process of the termination of an emergency and return to normal living conditions as outlined in [17] will need careful preparation. It is yet unclear how to achieve societal consensus about the conditions for returning to normal life.
– Transfer of information to the public is not a goal as such but essential for mitigating psychosocial impact. Information must be understandable and targeted for specific groups. This targeting may go down to single individuals who need specific attendance. Individual dose assessments based on suitable tools, namely incorporation measurements and dose reconstruction based on individual dwelling profiles, must be developed and made operational. Furthermore, such individual dose assessments have to be "translated" into understandable information for the population. This includes an assessment of the individual risk corresponding to the dose.
– Harmonization of protective measures is one key issue for credibility of EPR. Equal reference values and decision criteria among neighboring countries, cross border communication between decision-makers, and regular consultations in the preparation phase are prerequisites for a consistent protection strategy across borders.

References

1. Wirth E, Weiss W. Optimisation of the German Integrated Information and Measurement System (IMIS). Towards harmonisation of radiation protection in Europe: European IRPA Congress 2002, Florence, Italy, 8–11 October 2002.
2. Liktharev IA, et al. Thyroid cancer in the Ukraine. Nature. 1995;375:365.

3. Masson O, Baeza A, Bieringer J, Brudecki K, Bucci S, Cappai M, et al. Tracking of airborne radionuclides from the damaged Fukushima Daiichi nuclear reactors by European networks. Environ Sci Technol. 2011;45:7670–7.
4. Pittauerová D, Hettwig B, Fischer HW. Fukushima fallout in Northwest German environmental media. J Environ Radioact. 2011;102:877–80.
5. Zähringer M, et al. From dose rate to websites: making measurements accessible, understandable and helpful to the lay public. Radiat Prot Dosim. 2014;160:322–5. https://doi.org/10.1093/rpd/ncu204.
6. Steinhauser G, Brandl A, Johnson TE. Comparison of the Chernobyl and Fukushima nuclear accidents: a review of the environmental impacts. Sci Total Environ. 2014;470-471:800–17.
7. SSK. Weiterentwicklung des Notfallschutzes durch Umsetzen der Erfahrungen aus Fukushima, Empfehlung der Strahlenschutzkommission. Bonn; 2015.
8. Walter H, et al. RODOS-based simulation of potential accident scenarios for emergency response management in the vicinity of nuclear power plants. BfS-SCHR-60/16. urn:nbn:de:0221-2016091214084, Salzgitter, September 2016.
9. UNSCEAR. UNSCEAR 2013 report, volume I. Report to the General Assembly, Scientific Annex A: levels and effects of radiation exposure due to the nuclear accident after the 2011 great east-Japan earthquake and tsunami. 2013. www.unscear.org/docs/reports/2013/13-85418_Report_2013_Annex_A.pdf.
10. UNSCEAR. Exposures and effects of the Chernobyl accident (Annex J). New York: United Nations; 2000.
11. Tanigawa K, Hosoi Y, Hirohashi N, Iwasaki Y, Kamiya K. Loss of life after evacuation: lessons learned from the Fukushima accident. Lancet. 2012;379:889–91.
12. Yasumura S. Evacuation effect on excess mortality among institutionalized elderly after the Fukushima Daiichi nuclear power plant accident. Fukushima J Med Sci. 2014;60:192.
13. UNSCEAR. 2016. http://www.unscear.org/docs/publications/2016/UNSCEAR_WP_2016.pdf.
14. Vandecasteele C et al. Translation of the emergency planning zone to an intervention zone: a multidisciplinary approach improving common understanding and implementation of protective actions EUROSAFE Forum Brussels, 5–6 November 2012. 2012.
15. OECD. Towards an all-hazards approach to emergency preparedness and response: lessons learnt from non-nuclear events, OECD 2018, NEA No. 7308. 2018.
16. Safecast. 2018. https://blog.safecast.org/about/, version from 30.4.2018.
17. IAEA. Arrangements for the termination of a nuclear or radiological emergency, IAEA GSG-11, 2018. 2018.

MIX
Papier aus verantwortungsvollen Quellen
Paper from responsible sources
FSC® C105338

If you have any concerns about our products,
you can contact us on
ProductSafety@springernature.com

In case Publisher is established outside the EU,
the EU authorized representative is:
Springer Nature Customer Service Center GmbH
Europaplatz 3, 69115 Heidelberg, Germany

Printed by Libri Plureos GmbH
in Hamburg, Germany